P9-CBI-099

WIRELESS COMMUNICATIONS
Future Directions

THE KLUWER INTERNATIONAL SERIES
IN ENGINEERING AND COMPUTER SCIENCE

COMMUNICATIONS AND INFORMATION THEORY

Consulting Editor:
Robert Gallager

Other books in the series:

WIRELESS COMMUNICATIONS
Future Directions

edited

by

Jack M. Holtzman
David J. Goodman

Rutgers University
WINLAB
(Wireless Information Network Laboratory)

KLUWER ACADEMIC PUBLISHERS
Boston/Dordrecht/London

Distributors for North America:
Kluwer Academic Publishers
101 Philip Drive
Assinippi Park
Norwell, Massachusetts 02061 USA

Distributors for all other countries:
Kluwer Academic Publishers Group
Distribution Centre
Post Office Box 322
3300 AH Dordrecht, THE NETHERLANDS

Library of Congress Cataloging-in-Publication Data
Wireless communications : future directions / edited by Jack M.
    Holtzman, David J. Goodman.
        p. cm. -- (The Kluwer international series in engineering and
    computer science ; v. SECS 217. Communications and information
    theory)
        Based on the Third Workshop on Third Generation Wireless
    Information Networks, held in April 1992, Rutgers University.
        Includes bibliographical references and index.
        ISBN 0-7923-9316-3
        1. Wireless communication systems.   I. Holtzman, Jack M.
    II. Goodman, David J., 1939-   .   III. Series: Kluwer international
    series in engineering and computer science ; SECS 217.   IV. Series:
    Kluwer international series in engineering and computer science.
    Communications and information theory.
    TK5103 . 2 . W57   1993
    621 . 3845 ' 6--dc20   92-43366
                                                                    CIP

Copyright © 1993 by Kluwer Academic Publishers

All rights reserved. No part of this publication may be reproduced, stored in a retrieval
system or transmitted in any form or by any means, mechanical, photo-copying, recording,
or otherwise, without the prior written permission of the publisher, Kluwer Academic
Publishers, 101 Philip Drive, Assinippi Park, Norwell, Massachusetts 02061.

Printed on acid-free paper.

Printed in the United States of America

TABLE OF CONTENTS

vi

Preface

The past several years have been exciting for wireless communications. The public appetite for new services and equipment continues to grow. The Second Generation systems that have absorbed our attention during recent years will soon be commercial realities. In addition to these standard systems, we see an explosion of technical alternatives for meeting the demand for wireless communications. The debates about competing solutions to the same problem are a sign of the scientific and technical immaturity of our field. Here we have an application in search of technology rather than the reverse. This is a rare event in the information business.

Happily, there is a growing awareness that we can act now to prevent the technology shortage from becoming more acute at the end of this decade. By then, market size and user expectations will surpass the capabilities of today's emerging systems. Third Generation Wireless Information Networks will place even greater burdens on technology than their ancestors. To discuss these issues, Rutgers University WINLAB plays host to a series of Workshops on Third Generation Wireless Information Networks. The first one, in 1989, had the flavor of a gathering of committed enthusiasts of an interesting niche of telephony. Presentations and discussions centered on the problems of existing cellular systems and technical alternatives to alleviating them. Although the more distant future was the announced theme of the Workshop, it drew only a fraction of our attention. When WINLAB held its Second Workshop only sixteen months later, wireless communications had already gained recognition as a major trend in telecommunications. Debates were raging about competing second generation technologies, amid a growing awareness that there was much to be done to meet the needs of the long term future. Published versions of the presentations at that Workshop appear in the Kluwer book *"Third Generation Wireless Information Networks"* edited by Sanjiv Nanda and David Goodman.

The present book is based on the Third Workshop on Third Generation Wireless Information Networks, held in April 1992. By now, wireless information networks are recognized as key ingredients in the information services of the next century. Industry, government, academia, and the financial community all display intense interest in the subject and, in fact, conferences proliferate faster than new technology. In the beginning of this Workshop, we drew the attention of delegates to technical meetings centered on wireless communications to be held almost every month for the remainder of 1992! The delegates appreciated the format and future focus of this workshop. Attendance was limited to one hundred participants. There were no parallel discussions. Breaks and Workshop meals provided ample opportunity to debate issues raised in the formal presentations, as well as to discuss work in progress in centers throughout the world. The Workshop focused on the long term future. As a result, much of the work presented was intentionally preliminary in nature, but with the potential for future application. The Workshop generated substantial enthusiasm because of the ideas presented for discussion. We are pleased to present published versions of the Workshop presentations for the consideration of the entire wireless communications community.

The first section, **"Standards, Systems, Update,"** is a bridge from the present and near-future to the future farther out. Standards and systems issues for integrated voice-data systems are considered by C. A. Rypinski of Lace, Inc. The goal is a common air interface to serve personal portable computers and pocket telephones, automated warehouses and mobile robots and myriad other communicating devices. As an example of a Second Generation digital system, M. Kuramoto of NTT Radio Systems Laboratory discusses the Japanese Digital Cellular Standard. This provides a base of comparison and contrast to Third Generation Systems, such as the Universal Mobile Telecommunication System studied by RACE, an ambitious European project, described by C. Cengiz Evci of Alcatel Radiotelephone.

Section 2, **"System Issues,"** discusses two areas just beginning to get attention and which are critical to successfully implementing Third Generation networks. The first, by K. Meier-Hellstern, a WINLAB visiting scholar from AT&T Bell Labs, E. Alonso, a WINLAB visiting scholar from Ericsson, and D. O'Neil of BellSouth Enterprises, discusses the impact of mobility management upon Signaling System #7 (SS7). Although not originally designed to facilitate mobility management, SS7 is implicitly assumed capable of doing the job. The paper estimates the high signaling loads projected--it is a cautionary tale for those involved with intelligent networks. The paper by B. Gopinath and S. Das of WINLAB and D. Kurshan, a telecommunications consultant, discusses a new operating system for wireless networks. With software playing a crucial role in all large scale communications systems, it is essential to apply advanced computer science concepts to the new problems raised by mobility. Another area where computer science is needed is in organizing and optimizing databases for mobility management. We therefore included a contribution on that topic by T. Imielinski and B. R. Badrinath of the Rutgers University Computer Science Department and WINLAB.

Section 3, **"Alternative Approaches"** describes transmission media and signal processing techniques that have not been deployed in the cellular and cordless forerunners of the Third Generation wireless networks. J. M. Kahn, J. R. Barry, E. A. Lee, and D. G. Messerschmitt of the University of California-Berkeley consider a network capable of supporting 100 Mb/s infrared transmission links. Designed for wireless computer terminals, these transmission paths are a few meters in length. Far removed from this dazzling speed and small distances, J. A. Weitzen of UMASS-Lowell & Meteor Communication

Corporation, J. D. Larsen and R. S. Mawrey of Meteor Communication Corporation consider harnessing meteor bursts for vehicle tracking. The paper by J. Zander of the Royal Institute of Technology and K. Ahl of IDEON proposes a system with electronically steerable antennas to combine time and space multiplexing. Large potential capacity improvements result.

The theory of telephone traffic dates from the work of A. K. Erlang in the early years of this century. One of his enduring contributions is the Erlang B formula that has been pressed into service, appropriately or inappropriately, for a vast set of applications. Section 4, on **"New Traffic Approaches"**, demonstrates the need for new analytic tools to deal with problems unique to wireless communications. S. Nanda of AT&T Bell Labs derives new traffic models and applies them to determining cell sizes and consequent handoff rates in microcellular systems. S. Rappaport of SUNY-Stony Brook describes the effects on a network of groups of terminals, for example on a bus or a train, crossing a cell boundary simultaneously. E. H. Lipper and M. P. Rumsewicz of Bellcore use teletraffic theory to analyze co-channel interference in TDMA wireless networks.

Section 5 on **"Dynamic Channel Allocation"** continues a theme of the Second Workshop. The papers in this section all recognize that in advanced networks the static interference management techniques applied to cellular systems will not serve the needs of the Third Generation. With small cells, and low-altitude, low-power base stations, propagation conditions will be hard to predict. Traffic patterns will be even less stationary then they are today. Therefore it will be necessary for systems to adapt themselves to changing signal distributions and demands for service. The paper by G. J. Forschini, L. J. Cimini, Jr., and C.-L. I of AT&T Bell Labs emphasizes development of simple channel-allocation algorithms based only on local measurements. This results in a self-organizating system. J. Vucetic and D. D. Dimitrijevic of GTE Labs propose a hardware solution to adaptive channel allocation schemes. Analyzing dynamic channel allocation system schemes presents a formidable challenge--simulations are almost always used. Turning to analysis, J. Zander and H. Eriksson of Royal Institute of Technology provide asymptotic bounds on the performance of a class of DCA algorithms for large traffic loads. N. Bambos and G. J. Pottie of UCLA consider capacity maximization with power control. It is by no means clear at this point how all of these approaches compare, how they are related, and which (if any) will emerge as the methods of choice. The papers will provide a nice view of different viewpoints and approaches being taken. They serve as a base for further investigations.

The last section, **"PCN, Multimedia, CDMA"**, touches on several areas needing and getting attention. The paper by D. Raychaudhuri and N. Wilson of David Sarnoff Research Center discusses system design issues for multimedia

PCN's (Personal Communication Networks). The baseline definition of PCN is proposed to be broadened to include multimedia capabilities. The definition of PCN is itself a frequently debated topic and tying it into multimedia, another issue of great interest, deserves attention. The paper by R. Wyrwas, W. Zhang, M. J. Miller and R. Anjara of the University of South Australia continues the focus on multimedia. Possible access strategies using FDMA, TDMA, CDMA, and random packet access are presented and performance issues discussed. The paper by B. D. Woerner of Virginia Polytechnic Institute contributes to the continuing flow of papers on improved methods of evaluating CDMA. With CDMA competing with TDMA for many applications, CDMA performance evaluations are increasingly important.

We hope that you also will find the papers suggestive and stimulating.

Jack M. Holtzman
David J. Goodman
Rutgers University

Acknowledgements

We are grateful to the authors for their fine expositions, to Ms. Valerie Gould for invaluable help in preparation of this Volume, to Robert W. Holland, Jr. and Rose M. Luongo of Kluwer Academic Publishers for their support in the project.

ARCHITECTURE AND ACCESS METHOD ANALYSIS
FOR INTEGRATED VOICE-DATA SHORT-REACH RADIO SYSTEMS

Chandos A. Rypinski

LACE, Inc., Petaluma, CA 94954 USA

ABSTRACT

The reasoning is given for the choice of a single wide-band, maximum data rate channel used sequentially in small groups to resolve the frequency reuse problem. Channelization by means of frequency, time or code is found less flexible and adaptive for maximizing capacity and spectrum utilization. In addition, asynchronous use of a channel where there is a 5-step handshake sequence for access, with each transaction completed before the next use is started.

When there is need for a worst-case access delay control, as is essential for the use of a packet medium for virtual connections, then it is also necessary to limit the length of any allowed channel use.

A secondary and lower capacity means is available for direct peer-to-peer, and for spontaneous autonomous groups without infrastructure dependence.

OVERVIEW

The goal is a common air interface to serve personal-type portable computers and pocket telephones, automated warehouses and mobile robots and myriad other communicating devices.

Because it is also a goal to provide any desired degree of connectivity to other networks and to other members of the serving network, it is a corollary that this interface be transmittable over a variety of radio and optical mediums in space and in contained mediums using glass fiber, copper wires and cables of many kinds. The implementation of this must be economical of spectrum space, transmission medium, time, space and money.

The IEEE P802.11 Local Area Network Standards Subcommittee for Wireless Physical Medium and Access Method has undertaken to develop a Standard for this context.

SERVICES

The communication services to be provided are the same as those needed at a desktop or anywhere else. These are both packet with very high accuracy and moderate transmission delay and connections not only for compressed and uncompressed voice but also for all kinds of developing graphic and video applications. Bandwidth and destination must be available on demand at peak transfer rates for LAN at 10-20 Mbs and for connections up to 2 Mb/s.

METHODOLOGY

The focus of this paper is more on the reasons for the choice and the options considered and less on the detail of the offered proposals. There may be several ways to meet any one stated need. Of these only one will simultaneously satisfy a second and third constraint.

PHYSICAL MEDIUM CONSIDERATIONS

Because it is an "uncontained" medium, the radio or optical medium brings in many considerations that do not exist in wired systems. More attention is directed to a radio solution, but methods are avoided which could not be used equally with optical radiation. This consideration discourages use of medium specific functions like channel switching and signal strength measurement.

FREQUENCY REUSE

High spectrum efficiency requires that simultaneously usable transmitters be crowded as close to each other as they can be gotten. This problem is generic, and more difficult to define with shorter range. To get near 100% coverage for cellular telephone, 7 to 21 sets of frequencies are used with each set used for one coverage and including a setup channel.

Channelization

Channels can be provided by dividing time, frequency and code space to provide the 7 to 21 sets needed for continuous coverage. It will be asserted that: *one maximum width channel sequentially used as needed is a better way to obtain the reuse group function.*

Interference Limited Design

A corollary of high frequency reuse requires that coverage be limited by the aggregate level of interference from cochannel users--not the horizon or receiver sensitivity. For a complete system design there can be advantage in higher power and lower receiver sensitivity for milliwatt transmitters.

Channelization By Code Division

Code orthogonality in spread spectrum systems may be used as a means of separating low bandwidth users, base Stations in a reuse group or contiguous separately managed systems. These functions also can be done with time allocation or increased tolerance to a small amount of interference loss. The key advantage of channelization should be greater radio range for a given transmitter power. However, the processing gain obtained is often spent as an increase in tolerance to the noise generated by other simulataneous users.

FADE MARGIN

Very high fade margin is the primary over-design factor that can be reduced by diversity, redundancy and spectrum spreading. Fade margins sometimes result in 100 to 1000:1 overpowering of transmitters that are used in systems without anti-fade design. For high-rate data systems, failure to transfer correctly is often mistakenly attributed to insufficient signal level inducing further overpowering.

Large improvements in system capacity are obtained by minimizing power levels using diversity and appropriate modulation minimizing the effect and frequency of fades substantially below average signal level.

MOBILITY REQUIREMENTS

The non-arguable need for wireless is for user with no fixed location or who are actually moving while communicating. The importance of non-human Stations must be considered. However valuable, replacement of wiring for lower cost must be a secondary requirement.

Battery-operated Stations will certainly make up a large proportion of the applications and must be supported.

TOTAL RADIATED ENERGY

Minimization of aggregate as well as Station transmitter power is necessary for higher use density, and this is best done by keeping transmitters off except when performing a useful function.

SPREAD SPECTRUM RF TECHNOLOGY

The primary motivation for spectrum spreading is greater resistance against multipath fading. It is more beneficial when the separate paths may be resolved and used as a form of diversity. Reliance should not be placed on direct sequence spreading to mitigate the effect of narrowband interferers because of dynamic range limitations. If this advantage is attainable, it will depend on doing correlation and decoding at very low signal level.

There are classes of service for which code division channelization is not likely to be advantageous. These are wideband services only, mixed wideband and connection-type services or for bandwidth-on-demand connection-type services.

A model system plan and access protocol was developed on paper based on 9 code-division channels operated as a reuse group for data transfer and one channel for common setup. Based on access protocol and traffic capacity considerations, the subjective opinion was reached that it did not offer advantage over sequential use of the same code space at a higher transfer rate. The protocol function was far more complex and interlocked.

Use of code division to separate contiguous or overlapping systems with independent management may be technically feasible; but because of the low utilization of code space that results, this is not advantageous. The code space is better used to increase the transfer rate of a maximized single system.

Digital processing of received spread spectrum signals for data throughput rates above 10 Mb/s may be unusable for power drain and cost reasons.

Within these constraints and criteria, Barker type or other equally short symbol spreading is believed to be a suitable choice for maximizing the service capacity within a bounded area and bandwidth, and where a high transfer rate and demand-assigned bandwidth is part of the service requirement.

PROTOCOL CONSIDERATIONS
It is true that an access method cannot be designed without considering the limitations of the physical medium, and the reverse is also true. The access method is a vital part of maximizing spectrum utilization and service capacity.

HEADER-BASED PROCESSING
IEEE 802.3 and .5, ATM cells and fast packet are all prominent examples of protocols with header-based processing. The meaning and routing of a service or data-bearing frame at the receiving point is determined by the content of a header. This distinguishes against telephone PCM where the time position of a regularly recurring slot is an intermediate address form.

MINIMIZING LOGIC IN THE STATION
The telephone plant has historically made stations cheap and dumb and central equipment intelligent. The 802 standard local area networks of the computer community have made stations intelligent and made avoid the need for any central function. While avoidance of shared function was justified when these choices were made, there are other dominant considerations now.

It is essential to make Station design independent of the size of system, algorithms and various dimensional parameters which in the alternative can be implemented in the infrastructure. If this is not done, the design of the Station will be much more complex (not a decisive consideration), and it will be subject to updates and revision from time to time (much like software). A Station in which down-loading of software is the operating mode, also has a number of hardware complexity and operational disadvantages.

For a high-volume, standard-defined Station, it is extremely important that the function be stable over a period of years. for wireless systems, the computer community must accept common equipment and infrastructure.

SLEEP MODE IN STATIONS

An inactive state in Station equipment is obviously desirable for battery drain minimization, but there are also protocol advantages to a fine time structure for activating and de-activating Stations. Part of the justification for a polling function is that there must be a way for the connected networks to wake up sleeping stations to initiate a session.

STRUCTURED TIME USE

It is essential to use time in a planned rather than a fortuitous way. The algorithm for time use can and must be refined with time, and this is better done if it is entirely determined within the infrastructure function.

Without a central function, it is difficult to establish an orderly sequential use of the medium with queued traffic. Without queuing of traffic, it is not possible to approach 100% utilization without assuming that refused traffic disappears.

TRANSFER DELAY

In converting a 64 Kb/s connection to datagrams of 48 octets, there is a delay of 6 milliseconds while the samples are accumulated, and a further delay that is the worst-case for accumulated packets to get on the medium. The structural transfer delay for connections would be 9 milliseconds with 3 milliseconds arrival time tolerance. This value is much more than the 2 millisecond criteria of those in public network design, and much better than values necessary for use of highly coded speech.

ISOCHRONOUS OR ASYNCHRONOUS MEDIUM FORMAT

The choice is whether the medium should be isochronous with mapped in packets, or should it be packet with virtual connections. These alternatives are charactrized as either PHY or MAC multiplexing.

ISOCHRONOUS FRAMES

Isochronous frames are used by 802.6 and 802.9 with cells or packets mapped into assigned space along with other assignments of dedicated space for connection-type services. Slotted mediums with regularly recurring slot assignment are a form of channelization where the time position is an addressing mode. This allotment of a slot is a transaction that must be completed before the slot can be used. For handshake transactions, there must be a one frame length delay per round trip, and this is a one of the important reasons for the incompatibility of telecom and LAN methodology.

Adaptive Boundaries

With time division physical medium multiplexing (as in 802.9), there is an utilization when the boundary between isochronous and packet use is demand assigned. When the detail of this is implemented, the management of multiple connections and bandwidth-on-demand are substantial.

PHY Level Multiplexing

This is the case when the physical medium interface is common and the separate higher level stacks for connection and packet services are independent and multiplexed. With PHY multiplexing, only the medium transducer (and medium) is shared by voice and data services.

ASYNCHRONOUS PACKETS/CELLS

Packets or cells may be transmitted on a single digital channel provided that there is control that serializes incoming traffic. No information or destination coding is associated with time position, and the two-way negotiation to establish such slot definitions is avoided.

While cells can and are mapped into isochronous mediums (e.g. SONET and 802.6), they can also be managed into a serial bit streams.

High Channel Time Utilization

There must be queued traffic in order come close to 100% use of channel time. The delay to users is of the order of one or two average message lengths--a reason for an upper limit on packet length. Delay is more fairly distributed if users are served in order of arrival into queue (Erlang C blocking). Short service requests must have access to reserved time with priority ahead of data transfer, and this is one of the functions for which central control is advantageous.

8

MAC Level Multiplexing

MAC level multiplexing of the connection and packet services is above the MAC sublayer. The MAC (Medium Access Control) and the PHY layer are both common to isochronous and packet services.

Because of the special requirements of a radio PHY layer, the medium must be burst mode which is more easily adapted to packets than slots. A single MAC for the bursty medium is then more cost effective when it serves both virtual circuits and LAN services.

FUNCTION MOTIVATED CHOICES

Important choices depend upon detail aspects of the system function.

STATION POWER DRAIN

Minimization of battery drain requires that there be a defined sleep mode implemented in the MAC and system management. No Station function is active unless it is needed. Because of reduced circuit gain required and reduced frequency accuracy of components in the radio receiver, there is a further strong preference for wideband channels. Because of low duty cycle, transmitter power is less important than receiver functions to minimum battery drain.

COMMON CONTROL FOR MANY ACCESS-POINTS

To make the movement of a Station from one coverage to another a diversity event rather than a transfer between networks, it is important that external networks have a point of connection to the radio system which is independent of the particular Access-point from which a Station is served. Involving external networks with routing to particular and changing Access-points *can cause traffic storms* while connected networks attempt to find the new route for a moved Station.

CONSTRUCTIVE USE OF PATH REDUNDANCY

The transmission from a Station may be received correctly at more than one Access-point. It is valuable for the system to consider this an opportunity for selection diversity rather than a nuisance.

This concept is important to reducing losses from temporary path obstacles and to avoid need for the large fade margins often used to reduce this effect.

WIDE BANDWIDTH AND UNIFORM ENERGY DISTRIBUTION

It has been found that transmitted signal bandwidths of the order of 50 to 100 MHz at 2 GHz have significantly improved resistance to frequency selective fading when combined with appropriate modulation and processing. One of the conditions for obtaining this bandwidth is a near uniform spreading of energy across the available bandwidth.

It is also desirable to avoid peak energies much above the necessary values. Such peaks increase interference possibilities and the current requirements of amplifiers without offsetting benefits.

LOGIC DEFINED SEPARATION OF USER GROUPS

With independently managed and overlapping systems, is essential to replace channelized separation with a single wideband channel in which separation is obtained by logical identification or time sequencing of use.

With channelized separation, it is inevitable that many of the channels will be underutilized resulting in low utilization for the frequency assignment as a whole. With logical separation, the relative use of the shared channel is demand-assigned.

Logically separated systems then share capacity. To the degree that overlap can be minimized, the proportion of total capacity that must be shared is also minimized. It is an important requirement for the access method to anticipate the need for this function.

PROPOSAL HIGHLIGHTS--ARCHITECTURE

The intent of the definition of the air-interface is to provide the necessary functions with economical implementation. The cost of imagined hardware and installation configuration greatly affects judgments on technical methods.

INFRASTRUCTURE

Many in the computer community desire that each portable computer be able to communicate with others without dependence on infrastructure. This desire can be satisfied, but not at the same time providing a high degree of air-time utilization or a reach and path reliability that is comparable to the results obtainable with infrastructure. This is due in part to the possibility of directional antennas and diversity equipment in the Access-points and to much more favorable physical location for antennas at Access-points than can be assumed for portable computers.

Infrastructure is necessary for outside network access, and it is also required for large scale, high utilization and deterministic delay and capacity characteristics. Large areas can only be covered using infrastructure.

HUB CONTROLLER

The common controller, through transmission of *invitation-to-request* or *to-register* and of *poll*ing messages, manages the use of channel time and of sleep/wake mode in Stations. The Hub Controller is the preferred point of interconnection for external networks and the routing address for any Station that it serves regardless of which Access-point is providing coverage.

The Hub Controller is the residence of the algorithms that allocate capacity between Access-points and between different priority virtual circuit and data services. These may change with system size and evolutionary state.

The Hub Controller is also the residence of many management functions.

TOPOLOGY

The infrastructure includes a Hub Controller, usually located in the telecommunications wiring closet, which serves many wireless Access-points connected to it by pairs or point-to-point wireless medium.

Ordinary 24 gauge PVC insulated pairs, as defined in the EIA/TIA 568 "Commercial Building Wiring" Standard, transmit 16 Mb/s for distances of at least 100 meters from wiring closet to ceiling-mounted Access-point.

ACCESS-POINTS

The infrastructure point at which radio signals are received and transmitted are called Access-points. These are favorably located as compared with antennas on a portable computer. Each includes at least the radio or optical transducers to a low frequency or baseband waveform, and a wired interface for the link to the Hub Controller. Power is provided over the metallic path in the connecting pairs.

Radio Access-points are better installed in pairs in diagonally opposite corners of smaller rooms, or in corners of larger rooms as groups of four. The preferred coverage pattern covers a quadrant with maximum at the diagonal.

The Access-point may include selection diversity for multiple antennas. The protocol enables antenna selection decisions to be made before data is transferred during polls or setup.

AUTONOMOUS GROUPS

For relatively low channel usage or for many small autonomous systems within a common operating area, an alternate mode independent of infrastructure can be supported. Stations must support both modes, and the autonomous mode is inhibited when infrastructure is present.

No way is known to allow independent autonomous operation ("loose cannon") within the area of an infrastructure system that does not result in an unpredictable degree of capacity reduction.

PROPOSAL--AIR INTERFACE

The definition of the air-interface is the essence of what must be standardized and fully defined.

THE DATA SIGNAL

The transmitted signal is limited-length bursts of data using the full rate of the path--just a start-stop serial bit stream with embedded clock. The baseband form of this signal can be transmitted on telephone pairs and on guided or unguided optical mediums.

A medium data-rate on the channel should be at least 10 Mb/s. With a reuse group size of 4 sequentially-used, an average aggregate system capacity of 2.5 Mb/s per Access-point results. Since this capacity is dynamically allocacble between Access-points, the typical peak capacity is much higher.

Radio Frequency Modulation

The radio frequency carrier form is to be defined. If spread spectrum is selected for a medium transmission rate of 10 Mb/s or more, it is doubtful that more than 11 bits will be used as a spreading code. Unless phase coherent demodulation is used, more bandwidth will be used than 1 chip/Hz.

Operating Frequencies

The only immediate USA possibilities are the Part 15 ISM bands where there is 83.5 MHz at 2.4 GHz and 100+ MHz at 5.8 GHz. There is an imaginable future possibility of 40-140 MHz at 1.85-2.3 GHz.

If spread spectrum using the entire width of the higher ISM bands is considered, the chipping rate is unlikely to be above 100 Mchips/s. A more crowded situation would result from a narrower allocation in a new band.

It is an opinion that designers should do whatever they can do assuming a 70 MHz assigned bandwidth somewhere between 1.7 and 2.5 MHz. If FCC descisions are different, the results will have to be rescaled.

THE FRAME FORMAT

The format of the burst transmission on the medium is defined as a frame with header, payload and trailer. Burst frames are used at both Stations and Access-points.

The fields of the header are the bit and frame synchronization with start delimiter and the information necessary to process the payload. The payload may have internal formats, and it may have a length of zero. The trailer is the payload CRC and the end delimiter.

Frame Length Limit

There is no inherent fixed length, but a limit on the maximum size of payload to 288 octets or less is a good choice at signaling rates of 12 Mb/sec or higher. A high limit is more efficient for large size transfers, but a low limit provides fairer distribution of capacity at times of high service demand.

Segments

Long packets and virtual circuits are transmitted in segments of limited length. The present assumption is that the maximum length of packets and segments is limited to 288 octets matching needs for a 384 Kbits/s isochronous circuit. For reduced overhead, *it is desirable to use temporarily assigned and local short addressing for segments and virtual circuits.*

REGISTRATION OR SIGN-ON

For management reasons, Stations must register (same as sign-on in security system definitions) upon activation. The local short address is temporarily assigned upon registration, and is lost on de-registration. At registration, the Hub Controller determines the best Access-point to be used to reach that Station and adds that Station to the regular polling list. The span of a session is shorter than that of a sign-on.

Polling and the handshake functions are made more time-efficient by the availability of short addressing. If long addresses are considered payload, as they may be if passed to higher layers, this benefit will not necessarily appear in comparative efficiency calculations.

POLLING AND SLEEP MODE

Stations are periodically polled to determine that they remain reachable and to review the Access-point assignment. The poll may also be used to set sleep/active mode. The period between polls may be of the order of a second or less for less than 1% channel time use.

HANDSHAKE

The Station with a message to send must first hear an *invitation-to-request*, then *request* with full description of the service and destination requested.

The Station must hear a *grant* after which it can send the *packet* or *segment*. The destination responds with *acknowledge* or a *request-to-repeat* or nothing causing the originator to try again from the start.

No handshake is required for Access-point originated messages, because it may be assumed that Stations that are active and responding to a poll are always ready to receive. *The time lost in polling may be regained in avoiding handshake procedure on Access-point originated messages.*

Contention is possible for Station-originated requests, but not for data transfers originated at either Station or Access-point. Contention on requests is improbable, but a resolution by polling of currently active Stations is proposed. In this context, there are possibilities for optimizing the poll sequence and keeping the list short, which will greatly reduce the time taken to resolve this low probability contention event.

PEER-TO-PEER
Should the addressed destination Station be able to read the original transmission from the source Station, the repeat transmission from infrastructure may be canceled by prompt *ack*nowledgement from the destination. A delayed *ack*nowledgement is selectively used by the Hub Controller logic when, as a result of information obtained from the *poll* that there is a possibility of direct transfer.

OPERATION WITHOUT INFRASTRUCTURE
The same messages can be used for small groups of Stations to communicate without pre-arrangement or infrastructure using one of two methods: 1) Use of "Aloha" or CSMA enabling transmit from absence of signal with backoff, or 2) the first Station to become active emulates a subset of the infrastructure message functions.

The use of the autonomous mode should be locked out if infrastructure is present. The possibility of foreign stations operating within an existing infrastructure from which they are locked out must be dealt with in some way other than allowing them uncoordinated autonomous operation.

The infrastructure emulation is fairly simple when the management and large-scale functions are omitted. One station is charged with sending enable messages and polling those who are present. The protocol for the default access-manager can be in every Station selectively activated.

FIGURE

Shown in attached Figure 1 are sequence diagrams in approximate time scale for the main data transfer transactions. From top to bottom they are:

1) Normal short address Station-originate handshake and transfer
2) Normal long addresss Station-originate handshake and transfer
3) Auto *grant* Station-originate transfer
4) Normal short address Access-point originate transfer
5) Normal long address Access-point originate transfer

At a medium signaling rate of 12 Mb/s, 1½ octets per μsec are transferred. It is expected that the 4 μsec propagation and delay allowance shown will be reduced with system refinement.

The fields making up each message are approximately those described in contributions to the IEEE P802.11 Subcommittee (91-19, 91-80, 91-95). The acquisition time for a transfer is in detail field format, and is two octets plus a one octet start delimiter in this example.

DISCLAIMER

The intent is not display of originality and invention, though some such art might be present, but rather to collect those points of art on which a Committee might agree. The stated considerations and described methods are the Author's opinion only. The P802.11 Subcommittee has not made any decision on what access method(s) will be defined in the standard.

The Author wishes to acknowledge the great benefit and increase of understanding he has gotten from discussions with other members of the IEEE P802.11 Standards Subcommittee.

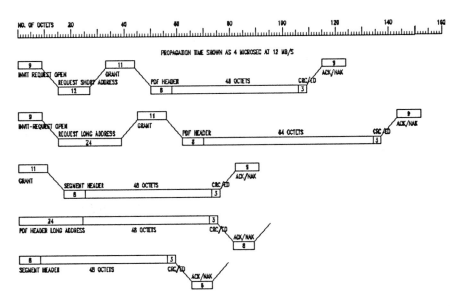

Figure 1   Message sequences with 5-step handshake for Station and Access-point originate data transfers

Development of a Digital Cellular System Based on the Japanese Standard

Minoru Kuramoto
R&D Department, NTT Mobile Communications Network, Inc.
1-2356 Take, Yokosuka-shi, Kanagawa-ken, 238-03 Japan.

Abstract

The design concept, new services, and technologies of the NTT digital cellular system based on the Japanese Digital Cellular (JDC) standards are described. The digital system offers high quality facsimile/data as well as the conventional voice service. Compact and long battery life, portable digital telephones will be introduced. System capacity can be increased by using diversity, microcells, low-bit-rate speech coding, and new frequency (800/1500 MHz) bands.

1. Introduction

Public cellular mobile telephone services have changed greatly in both the quality of service and the number of subscribers in recent years. NTT started an analog cellular service in December 1979 in Japan using the 800 MHz frequency band. The introduction of market competition and the development of portable telephones quickly led to an explosive growth in the number of cellular telephone users. The number of subscribers will approach 1.4 million in March, 1992 and is approaching the capacity limit of the current analog cellular systems. The number of subscribers is expected to continue to accelerate; we anticipate 10 million subscribers by the year 2000.

The shortage of available frequencies is an obstacle preventing services expansion not only in Japan but in North America and Europe as well. New cellular systems must be introduced that offer higher capacity and a capability for providing a variety of new services. Digital cellular radio systems are now being developed throughout the world to meet these requirements. In this paper, the basic specifications for a unified Japanese digital cellular (JDC) standard [1] are described together with an outline of the NTT digital cellular system. Also presented are future development that will support new services.

2. JDC standard

The major service in current analog cellular systems has been voice communication. Recently, however, there is growing demand for a variety of new, high quality services. Currently, there are three different analog cellular systems in Japan. However, the "roaming" ability to access any cellular network throughout the country is essential from the user's point of view. A fully digital, unified air-interface between the subscriber unit and the base station is indispensable for this capability. The Digital Mobile Telephone System Subommittee of the Telecommunications Technology Council has studied the technical conditions for the digital cellular system. Based on this report, the fully digital, JDC standard was established by the Research & Developemnt Center for Radio Systems (RCR) in April, 1991.

2.1. Common Air-Interface
Since the current NTT analog cellular system has an extremely high spectrum efficiency (6.25 kHz separation), the next generation cellular system is required to offer superior performance. Digital technologies offer such potential advantages as greater capacity, lower equipment costs, new services and upgraded portable handsets. Fig. 1 summarizes these advantages and the associated technologies. Efficient spectrum utilization is achieved by employing a narrow band digital modulation technique and a robust low-bit-rate speech CODEC. Furthermore, enhanced radio link control will make possible smaller cell layouts. Base station installation costs will decrease with the employment of TDMA. An OSI based layered signaling structure and functional module will provide the needed to allow new services to be added as required. The intensive application of LSIs will result in compact and inexpensive portable sets.

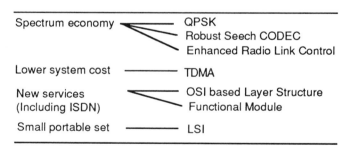

Fig.1 Advantages and Applied Technologies

The basic radio channel design defined as the JDC standard is compared in Table. 1 with those of the North American TDMA digital system (NADC) and European GSM system. The JDC systems use 3-channel TDMA. Two frequency bands are reserved: the 800 MHz band with 130 MHz of duplex separation and the 1.5 GHz band with 48 MHz of duplex separation. The 800 MHz band will be used first. The 1.5 GHz band will be a later use. The modulation scheme is spectrum efficient $\pi/4$-QPSK so that the interleaved carrier spacing is 25 kHz. The speech CODEC was selected based on evaluations of the quality, complexity, and delay of several speech CODECs. The selected

Table 1. Basic Specifications

	JDC	NADC	GSM
Frequency Band Up link	940 - 956 MHz 1429 - 1441 MHz 1453 - 1465 MHz	824 - 849 MHz	890 - 915 MHz
Down link	810 - 826 MHz 1477 - 1489 MHz 1501 - 1513 MHz	869 - 894 MHz	935 - 960 MHz
Carrier Spacing	25 kHz Interleaving	30 kHz Interleaving	200 kHz Interleaving
Modulation	$\pi/4$ QPSK		GMSK
Multiple Access	3(6)ch TDMA		8(16)ch TDMA
Carrier Bitrate (CODEC)	42 kb/s (VSELP 11.2 kb/s)	48.6 kb/s (VSELP 13 kb/s)	270 kb/s (RPE-LTP 22.8 kb/s)
Equalizer	Option	Mandatory	Mandatory
Diversity	Option	Option	Option
Frequency Hopping	——	——	Option

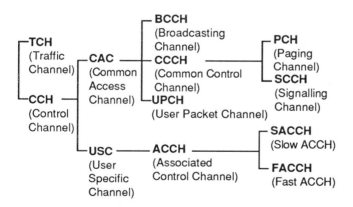

Fig. 2 Logical Channel Structure

CODEC uses 11.2 kbps VSELP (Vector Sum Excited Linear Prediction) including channel coding.

2.2. Channel Structure
(1) Logical channel
The logical channel structure is shown in Fig. 2. The channel types are divided into two: one is the traffic channel (TCH) for user information transmission and the other is the control channel (CCH).

There are two types of CCH: a common access channel (CAC) shared by plural users, and a user specific channel (USC) dedicated to a respective user. The CAC is further divided into three. The *broadcast control channel (BCCH)* provides the mobile stations with system information which contains the mobile station-related information such as the maximum transmission power of the mobile station, location identity codes to register the user's location, and the information related to the control channel structure, such as the number of control channels available in the cell. The *common control channel (CCCH)* is a point-to-multi-point two-way channel used for signalling information transmission, and comprises the paging channel (PCH) and signalling control channel (SCCH). PCH, which is broadcast from multiple cell sites within a paging area, transmits paging information to mobile stations. SCCH is used to communicate, between the network and the mobile station, signalling information other than the paging message. The *user packet channel (UPCH)* is a point-to-multi-point two-way channel used to transmit user packet data.

The USC is the control channel associated with the TCH and defines the associated control channel (ACCH) used to transmit the signalling data required during communications. There are two control channel types; SACCH (slow ACCH) and FACCH (fast ACCH). The SACCH has a slow data transmission rate, while the FACCH is a high data transmission rate channel using the TCH bits.

A radio-channel house-keeping channel (RCH) is also provided. It is a layer 1 channel without the layered signalling protocol, and is used for control functions that require real-time responses such as transmission power control and state notification.

(2) Physical channel mapping
TCH is defined with a six slot TDMA format; however, initially, two of them (therefore 3 channel TDMA) will be used by one user until a half-rate speech CODEC becomes avialable. The common access channel is only defined as a full-rate format. One of these TDMA slots is called the "physical channel," and their structures are shown in Fig. 3.

The CAC physical channel mapping is shown in Fig. 4. The BCCH, PCH, and SCCH frames can be recognized by the position in the superframe structure.

2.3. Layered Signalling Structure
A 3-layered signalling structure based on the OSI model is adopted to allow independent development of the transmission methods and control signals, as shown in Fig. 5.

(a) TCH/ACCH

Upward:

| R 4 | P 2 | TCH(FACCH) 112 | SW 20 | CC 8 | SF 1 | SACCH (RCH) 15 | TCH(FACCH) 112 | G 6 |

Downward:

| R 4 | P 2 | TCH(FACCH) 112 | SW 20 | CC 8 | SF 1 | SACCH (RCH) 21 | TCH(FACCH) 112 |

(b) CAC

Upward — First unit:

| R 4 | P 48 | CAC 66 | SW 20 | CC 8 | CAC 116 | G 18 |

Second unit:

| R 4 | P 2 | CAC 112 | SW 20 | CC 8 | CAC 116 | G 18 |

Downward:

| R 4 | P 2 | CAC 112 | SW 20 | CC 8 | CAC 112 | E 22 |

G : Guard Time
R : Ramp Time
P : Preamble
SW : Synchronization Word
CC : Color Code

SACCH : SACCH bits
FACCH : FACCH bits
RCH : House keeping bits
SF : Steal Flag
CAC : CAC bits
E : Collision control bits

Fig. 3 Physical Channel Structure

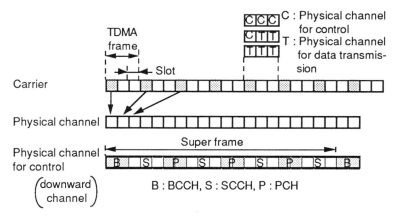

Fig. 4 Mapping of Logical Channels on a Physical Channel

Layer 2 consists of the address and control parts. The address part allows a base station and plural mobile stations to exchange information over the CCCH. The control part initiates the retransmission of signals.

The layer 3 is divided into three modules; call control (CC), mobility management (MM), and radio transmission management (RT). Messages in the three modules are transmitted in a single layer 2 packet with the help of the layer 3 common platform, thereby improving radio frequency utilization. The function of RT/MM is to control the setup and maintenance of the radio links, handover and so on. The CC function is based on the ISDN user-network interface layer 3 (I.451) in order to guarantee the expansion of services and interconnection with ISDN networks.

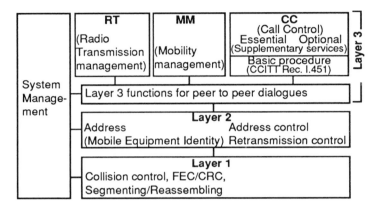

Fig. 5 Layered signalling structure

3. NTT Digital Cellular System

The JDC standard defines only the air interface. The design of the network is left to the respective operators. In this section, the network configuration and system configuration of the NTT digital cellular system are described.

3.1 Network
Digital cellular networks should be designed to fulfill three requirements:
(1) Connectivity with fixed networks (PSTN, PSPDN, and ISDN)
(2) Mobility (roaming) of the mobile stations between different cellular network operators
(3) Connection of the cellular and fixed networks with a unified interface

Fig. 6 shows such a cellular network. It consists of the mobile communications control station (MCC), the base station (BS), and the mobile staion (MS). The MCC comprises a gateway MCC (G-MCC) providing a toll switch and gateway functions to the fixed network, a visited MCC (V-MCC) providing call connection and bearer connection functions, and a home location register (HLR) in which the mobile station identification numbers and subscriber locations are registered. CCITT No. 7 signalling is used for control signal transmission between MCCs.

All connections with the fixed network are established through the interworking gateway switch (IGS) using the No. 7 signalling system ISUP (ISDN User Part). This permits, except for certain limitations, the wide range of ISDN services to be provided over the cellular system. In addition, any cellular network can be easily interconnected.

When the MS moves from its home network to another cellular network (roaming), it is necessary to track the connection between networks after confirming that the roaming MS is an authentic MS. To provide the NTT system with this connection tracking ability, a gateway LR (GLR) is provided to temporarily store the data of MSs entering the system from other networks. For a call from the fixed network to the MS, the HLR is first accessed to transfer the subscriber data to GLR in the terminating cellular network, and the MCC then pages the MS.

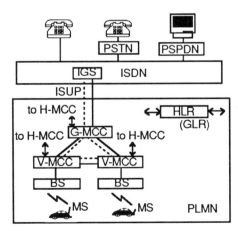

G-MCC : Gateway-Mobile Control Center
V-MCC : Visited-Mobile Control Center
H-LR : HomeRocation Registor
IGS : Interworking Gate Switch
BS : Base Station
MS : Mobile Station

Fig. 6 Network configuration of the digital cellular system

3.2 System Configuration

The system configuration of the NTT digital cellular network is shown in Fig. 7. The MCC comprises the mobile control and switching system (MCX) and the speech processing equipment (SPE) for transcoding between 64 kbps m-law PCM and 11.2 kbps VSELP (including channel coding) for speech. The three traffic channel

24

information (corresponds to one radio carrier) can be transmitted using a single 64 kb/s channel between MCC-BS communications links.

The BS comprises the base station control equipment (BCE), base station modulation/demodulation equipment (MDE), and base station transmitter-amplifier equipment (AMP). The BCE can also be installed in the MCC. In addition, by using a multiple carrier amplifier, the base station becomes both smaller and less expensive than the existing analog stations. In urban areas, the space available for base station equipment may be very small and thus very compact equipment is required. Using optical fiber links, a base station can be constructed with only a transmitter and a receiver amplifier.

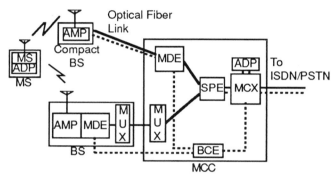

Fig. 7   System Configuration

4. System Features

The major features of the NTT systems are shown in Fig. 8. The NTT system aims to realize high capacity with dynamic channel allocation, small portable phones with antenna diversity, voice operated transmission (VOX) and efficient linear power amplifiers. Lower system costs are achieved through the use of multiple carrier power amplifiers. High quality telematique services are provided, and the ISDN services will be offered through the development of an ISDN protocol.

4.1 Increased Capacity
The capacity can be estimated and compared with analog systems considering the frequency and space utilization factors of the radio spectrum. The frequency bandwidth per traffic channel is 8.3 kHz (= 25 kHz/3 interleaved channels) for the digital system, and 6.25 kHz (interleaved channels) for the NTT high capacity analog system. Thus, the ratio h_f of the frequency utilization factors is 0.75 (6.25/8.3). Therefore, the frequency factor of the digital system is slightly less

than that of the analog systems. The frequency reuse pattern, which is a measure of space utilization factor, is a 7-site × 3-sector in current analog systems. Digital systems can realize (a) robustness against co- and adjacent-channel interference, and (b) reduced interference by the site diversity effect due to the precise cell determination using MAHO (mobile assisted hand over). As a result, a 4-site × 3-sector cell layout can be realized. The ratio of h_S of space utilization factors is increased 1.7 times (=7/4). Consequently, the digital system has initially, 30 % (=6.25/8.3 × 7/4) larger capacity than the analog system. Introduction of half-rate speech coding, dynamic channel assignment, and micro cells can further increase the digital capacity.

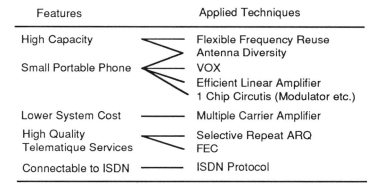

Fig. 8 NTT System Features

4.2 Improved Radio Channel Quality

(1) Diversity reception
To combat multipath fading effects, diversity reception is adopted. The base station transceivers use postdetection selection diversity reception. At the mobile stations, more simpler antenna selection diversity using only one receiver is used; the antenna is selected by measuring the received signal levels of the two antennas before reception of the allocated time slot. The measured average BERs under Rayleigh fading are shown in Fig. 9. The use of diversity reception can achieve BERs under than 10^{-2} if the rms delay spread t_{rms} is less than 8 ms. Combined with base station antenna beam tilting, we expect that without the use of complicated equalizers, delay spead effects can be significantly reduced.

(2) Speech quality
The combined use of signal scrambling and encryption offers an extremely high level of security and confidentiality without any deterioration of speech quality. Furthermore, the digital system can

26

provide almost constant quality up to a bit error rate of about 1~2% and thus within a entire cell, the quality is kept constant. Analog cellular speech quality on the other hand, gradually deteriorates as the MS moves away from the BS.

(3) Error free data transmission
An adapter is provided at MS and MCC for converting between voice-band modem signal and 11.2 kbps JDC-TCH data. Error free facsimile signal transmission is achieved with the use of the high-throughput ARQ transmission scheme. The widely used MNP modem protocol is also provided for data communications.

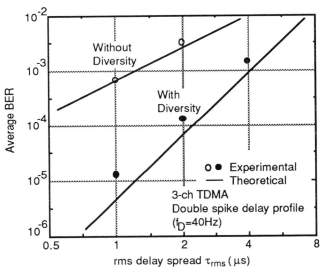

Fig. 9 BER Performance

4.3 Smaller and More Economical Equipment
The base station equipment is approximately one-fifth the size and significantly less expensive than the analog system equipment due to the use of 3-channel TDMA , multiple carrier amplification, and centralization of the control equipment. Compact portable units can be realized by massively employing LSI and VLSI devices, reducing power consumption with voice oprated transmission (VOX) and substituting diversity reception for multipath channel equalizer.

5. Future Evolution

The next evolution of digital cellular services is discussed in this section (see Fig. 10). Digital cellular services will consist of circuit exchange services, packet exchange services, cellular original services and universal personal telecommunications (UPT) services. In circuit exchange services, voice communication will be the main service in both digital and analog systems. A half-rate speech CODEC with 5.6 kb/s will be introduced within a few years. Unrestricted digital information services (of the order of 2-8 kbps and 64 kbps) will be introduced in the near future. Packet exchange services using UPCH, CCCH or ACCH are also items for the future. Cellular original services such as short message and location data are also important in the near future. UPT services are the keys to the next evolution of digital cellular services. Much emphasis should be placed on the total research and development of UPT networks and terminals. In the following, some of the services to be introduced in the near future are described.

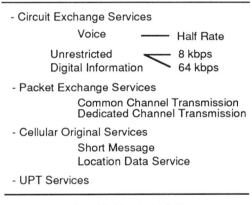

Fig. 10 Next Evolution

(1) Short message service
By using random access methods in UPCH, CCCH or ACCH, a number of messages can be transmitted and received efficiently. There are two communication modes in this service.

In one mode, the ACCH is dedicated when TCH is used for conversation (or facsimile/data communication). In the other mode, several terminals access the same UPCH or CCCH. Functions peculiar to the short message service may be provided by a short message handling module (SHM), which has a gateway function for PSTN, a message handling function, a message charging function and so on.

(2) ISDN services
ISDN services, now available in the fixed system, can be extended to the cellular system, such as 64kbps unrestricted digital information (UDI) and packet-mode bearer services. 64kbps UDI requires one or

two 64kbps channels (B channel) and 16kbps channel (D channel) on the air interface. For this service, new TCHs must be defined on the JDC air interface.

(3) Packet service
The JDC standard has already defined the UPCH channel for the packet service. This service has the two communication modes as described for the short message service. It is necessary to introduce a mobile packet handling module (MPHM) which has the CCITT standard protocol X.75 for connection to ISDN and/or PSPDN and a packet mode protocol such as the CCITT standard protocol X.31 for an interface between terminal equipment (TE) and MS.

6. Conclusion

The design concept of a digital cellular system based on the JDC standard was presented. In the 800 MHz band, only 2×16 MHz is available for all digital cellular operators. Considering the more than 50% annual growth rate experienced by current analog cellular systems, saturation of cellular capacity will be unavoidable with this 2×16 MHz band by 1994-5. In order to cope with this, our next targets are to use the 1.5 GHz frequency band, to increase system capacity, and to create enhanced services.

Development of the NTT digital cellular system based on JDC standard will be completed so that commercial service can begin in 1992. In Japan, in addition to the JDC standard, a unified standard for the Personal Handy Phone (second generation cordless telephone) is under discussion and will be established this year. It will adopt 4-channel TDMA/TDD with 384 kbps. From this March, validation tests for the air-interface will be conducted.

For the realization of FPLMTS, a more frequency-efficient access scheme is necessary. This goal will be attacked with more intensive research & development activities.

References
[1] "Digital cellular telecommunication system RCR standard," Research & Development Center for Radio Systems, STD-27, Apr. 1991.

PAN-EUROPEAN PROJECT FOR THIRD GENERATION WIRELESS COMMUNICATIONS

Cengiz EVCI and Vinod KUMAR
ALCATEL RADIOTELEPHONE
Paris-FRANCE

Abstract
Universal Mobile Telecommunication System (UMTS) is a projected European standard of the CCIR defined Future Public Land Mobile Telecommunication Systems (FPLMTS) for third generation mobile communications. RACE Mobile project is studying issues related to UMTS. RACE is the European Commission's programme on Research and development in Advanced Communications in Europe. The main purpose of this programme is to introduce an Integrated Broadband Communications Network (IBCN) so that it provides an universal standard telecommunications system across Europe, with the capacity to meet demand in the early 21st. century.

The purpose of this paper is to make the reader familiar with the RACE Mobile Project which is partially funded by Commission of European Community (CEC). Further, the paper describes many techniques investigated within the project with respect to UMTS requirements.

1. INTRODUCTION

The ultimate aim of mobile communication lies in ensuring an exchange of every information with *"anyone"*, *"anywhere"* and *"anytime"*, at low costs using handy devices. These facts resulted in a rapid growth in the demand for mobile communications and consequently, they have led the industry into intense research and development efforts towards a new generation mobile systems.

Today, in the European mobile market, four main types of mobile systems are dominant: cellular (GSM-Global System for Mobile Communications or Pan-European Second Generation Digital Mobile System), cordless systems (DECT-Digital European Cordless Telecomms.), paging systems (ERMES-European Radio Messaging System) and private mobile radio (PMR-largely based on the UK MPT-1327 standard). Moreover, in each of these four areas developments are taking place towards European systems,i.e. European Standards are being drafted.

These four types of systems are not compatible, i.e. a car telephone can not be used on a cordless system. By the turn of century, all these incompatible systems will have stimulated the mass market for mobile communications. By then however, the standards will have been stretched to their full capacity and will fail to fully satisfy user requirements and demands. It will be time for a third generation system : universal, multi-function, digital system based on ideas of the 1990's and on technology of the turn of century. The aim of the UMTS is to integrate the four types of mobile communication into one system with the pocket telephone as principal use.

Market estimates indicate that around year 2010 some 150 million pocket phones would be used in Europe. The UMTS is therefore to be designed to cope with this enormous mobile traffic [1]. Consequently, research and technology aspects to meet that demand in the early next century are real challenges. The important objectives of UMTS are mainly three fold. First one is related to the universality- terminal will have to work almost anywhere with high traffic density. Second, the pocket phones with long battery life should provide a high service quality- at least as good as today's fixed network quality. Third, they must have multi-service capability for voice, data, text and image communication. Research work on wide range of technological issues is being undertaken in the RACE Mobile project, under the umbrella of the CEC-RACE programme. This project sees future telecommunication services as being provided by UMTS. In order to meet the objectives of UMTS, it is believed that it will include micro-cellular techniques, fast and reliable handovers, adaptive frequency planning methods, effective modulation and voice coding [2,3].

The aim of this contribution is to report on the work of UMTS research carried out under RACE from its start up to now. After a brief history of the work organisation in the first chapter, we present the general requirements of UMTS, followed by the concept of UMTS and services. This will make the reader familiar with the UMTS. Then, the work done for UMTS to meet these requirements and the corresponding technical choices will be addressed.

In order to achieve the above goal, we initially focus on the fixed network area, and highlight the major results. Later, the work for radio interface design will be described and this will comprise some specific topics like digital coding of speech for UMTS requirements. Further, the system group activities with special emphasis on system framework for UMTS design will be mentioned. Finally, under the title of miscellaneous issues, the review of some work done on techno-economic of the UMTS cellular network, implementation and the contribution of Mobile Broadband Systems (MBS) group to RACE Mobile project are given.

2. HISTORICAL BACKGROUND

In beginning of 1988, manufacturers, operators, universities from CEC and EFTA (European Free Trade Association) in Europe came together to create a consortium for RACE-Mobile project on a third generation wireless communication system which will be operational in the twenty first century [4]. This project (RACE1) with two phases I&II of three years (1988-1990) and two years (1991-1992) duration respectively is being followed by RACE2 of three years (1992-1994) duration. The first phase (phase I) of the work was carried out with eight core areas. These core tasks have been divided into many workpackages. Although considerable interaction between the workpackages is required, great care has been taken to avoid duplication. The phase I plan was as follows:

- **Fixed Network (FN)** : to introduce the functional specification of the fixed network for the mobile communications,

- **Radio Bearer (RB)** : to study the channel behaviour and equalisation, modulation and channel coding,

- **Mobile System&Services (MSS)** : to study the complete system and to prepare reference documents, called system description document of the UMTS,

- **Cellular Coverage (CC)** : to study handover techniques and microcellular infrastructure,

- **Channel Management (CM)** : to produce an air-interface specification,

- **Signal Processing (SP)** : to investigate the implementation aspects for the solutions proposed in other core tasks. The speech codec is related to this task as well.

- **Other Technology (OT)** : to study antenna systems and the architectures of radio frequency sections of terminals,

- **Mobile Broadband Systems (MBS)** : to search for the techniques to provide services requiring bit-rates varying from 1 to 155 Mb/s.

The first three years of RACE1 were spent by the consortium companies to perform an open ended research work in various fields related to mobile communications. This phase generated basic technical conclusions so that a "**top-down**" system design work could be undertaken during phase II of RACE1. During second phase, core areas FN, RB (replaced by MEC-Modulation, Equalisation and Coding), CC, CM and MBS were retained. In addition, **Systems Group (SG)** was introduced. The general objectives of this new group are as follows:

- to draw up a realistic requirements specification for a third generation systems,

- to collate, clarify and draw together all technical results in the project with a overall top-down system view,
- to assess the system framework against the requirements,
- to contribute technical information to relevant standardisation and other bodies.

Most of the conclusions of the RACE Mobile project are contained in the formal "deliverable" documents presented to the CEC. These documents are available only to members of the RACE Industrial Consortium (RIC). However, many results are also available in the public domain via papers presented to various conferences and standardisation bodies. Only these latter documents are referenced in this paper.

At this point, certain events in parallel to RACE1 are worth consideration :

- CCIR/CCITT interest in Personal Communications and advancement of FPLMTS,
- Momentum gained by CDMA applications to mobile communications [5],
- Adoption of GSM standard by U.K. Personal Communication Network (PCN) operators under the name DCS1800 and start of design in early 1990,
- Creation of ETSI (European Telecommunications Standards Institute) Sub-Technical Committee (SMG5) on UMTS which will work towards a standard for third generation of mobile systems by 1998,
- Outcome of WARC'92 as far as the frequency of UMTS is concerned.

In the coming chapters, after an introduction of UMTS concept, the technical achievements in different study areas for mobile communication system design are reported.

3. THE CONCEPT OF UMTS

A UMTS is intended to satisfy the demands for next generation mobile systems and will integrate the functionality of the present mobile systems into a single replacement system. This chapter is devoted to the UMTS concept and can be analysed under three sub-sections, namely, requirements, concept and services.

Requirements of UMTS

The requirements for UMTS [3] can be summed up as follows:

- Spectrally efficient communication system by which every user can exchange information with anyone, anywhere, at anytime.

- High capacity to support high market penetration,
- Support for a wide range of services,
- Indoor/outdoor coverage,
- Light, pocket size, low cost handset with long autonomy,
- High service quality: at least quality of today's fixed networks,
- Access to different services possible from single terminal type,
- Communication security
- Fair charging.

UMTS Concept

The UMTS should provide a wide range of services to a high number of users and services should be available to a wide range of radio terminal types. UMTS is based on three fundamental concepts [2,3,6]:

- A sufficiently standardised environment such that mobile communication is viable in almost any location,
- A wide range of services including high quality speech, high bit-rate services and subsets of Narrowband (N-ISDN) and Broadband ISDN (B-ISDN) services (IBCN is the concept, B-ISDN is the implementation of IBCN),
- A small personal communicator which is small enough to be carried comfortably in the pocket and of low cost.

The key to realise a system to support a pocket telephone usable in all environments is flexibility in the air-interface. The constraints-requirements of indoor and outdoor are so different that no single choice of system parameters is optimum for both.

The available spectrum for UMTS will be limited. In order to cover the different environments for various densities of mobile users, a number of cell types and sizes are required. For example, small cells offer high frequency reuse and thus high capacity but can not be economically used in areas of sparse population. The cell types [2,7] defined are **macro-cell** (rural and suburban areas), **micro-cell** (urban areas and city centres), **highway cell** (highways outside urban areas), **pico-cell** (indoors and e.g., on a train) and **overlaying macro-cells** (a large cells overlaying many smaller cells) as shown in Figure 1.

The UMTS will consist of both public and private networks combined in such a way that the customer can roam (between and during calls), be located and billed correctly. UMTS will thus be comprised of many operators.

34

The UMTS should, where possible, support the services provided by other networks, such as ISDN, B-ISDN and private networks for example, subject to political, technical and commercial constraints.

Considering all these requirements, the UMTS can be visualised as in Figure 2, one communicator for all environments and universal possibility to roam between the different networks using the same terminal [1-3,7-8]. It is clear from this figure that UMTS is seen to be a multi-environment system, be a multi-operator system, include both public and private sub-networks in order to provide in the four environments shown in figure (the home-DCPN, business-BCPN, vehicle-MCPN and public environment) and allow inter-operation with the other existing or future telecommunications networks, both fixed and mobile, and allow also integration into the B-ISDN.

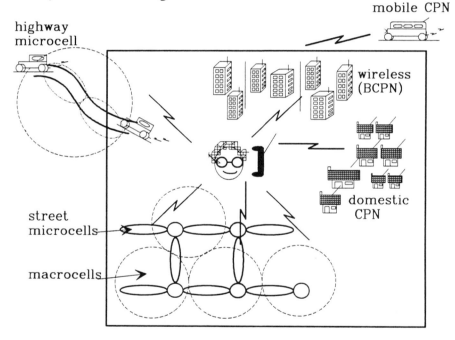

Figure 1 : UMTS Sub-networks and cell sizes [2]

In the home, business and mobile environments, the services are to be provided by private DCPN, private BCPN and public MCPN (for example on a bus, train or boat), respectively.

Ultimately, the UMTS backbone network shall be B-ISDN, however, ISDN may be utilised during the evolutionary phase in accordance with its capabilities.

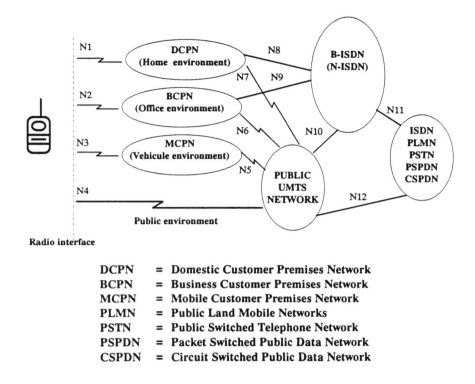

DCPN	= Domestic Customer Premises Network
BCPN	= Business Customer Premises Network
MCPN	= Mobile Customer Premises Network
PLMN	= Public Land Mobile Networks
PSTN	= Public Switched Telephone Network
PSPDN	= Packet Switched Public Data Network
CSPDN	= Circuit Switched Public Data Network

Figure 2 : The concept of UMTS : the provision of mobile services in different environments

The interfaces between the different UMTS subnetworks as well as between UMTS and other telecommunication networks are indicated by reference points N1-N12 in Figure 2. The characteristics of N1-N12 are :

- **N1-N4** : Radio interfaces between a user terminal a radio access ports of UMTS, whether in the home, office, vehicle or outdoors.

- **N5** : Radio interface to cope with a number of UMTS users within a single MCPN sub-network (on a bus, train or boat).

- **N6-N7** : Radio interface that may be used as a replacement for a wired local loop to connect a BCPN or a DCPN, respectively.

- **N8-N9** :The user-network interfaces between the fixed telecommunication network and DCPN and BCPN, respectively.

- **N10** : Network to network interface that connects UMTS-PLMN to a backbone fixed network.

- **N11** : When the UMTS is integrated into a backbone network the interworking to other networks is through that backbone network.
- **N12** : The UMTS stand-alone network may interwork with other networks. Interworking functions may be provided by the UMTS or other network.

UMTS Services

A UMTS will support a wide variety of services [3,6]. UMTS aims at the provision of at least all GSM, DECT and N-ISDN services where ever possible. It should be noted that in certain areas such as rural area, the system capacity could be too low to offer all of these services. Next to this set of minimum services (that will be provided by an operator in case of sufficient system capacity), wideband services can be considered in high capacity areas such as city centres. In the field of teleservices additional UMTS services like video telephony, wide-area paging, video-audio (multi-media) information transmission, image-document-data retrieval, audio-speech retrieval can be considered. As far as bearer services are concerned, next to bearer services already defined for ISDN and GSM, circuit-mode bearer services like 8, 16 and 32 kb/s unrestricted could be taken into consideration along with higher rate packet-mode bearers. Most ISDN and GSM supplementary services are expected to be supported by UMTS.

4. NETWORK ASPECTS OF UMTS

The mainstream activities of the fixed network group of this project were aimed at investigating network architectures and functional models for UMTS, describing mobility procedures within the network, and considering a model for storage, maintenance and access of data required to operate the system [2,6-7]. Some results are highlighted in the following sub-sections.

Network Architecture

Figure 3 gives an overview of a possible network architecture for UMTS, showing network entities and interfaces in public environment. The main UMTS specific network entities distinguished are the Mobile Terminal Equipment (TE_m), Base Station (BS), Mobile Control Node (MCN), Information Storage Node (ISN). The Network Terminations (NT1), Line Terminations (LT) and Switch (SW), as well as the S, T, U and V interfaces have been chosen in line with B-ISDN to favour integration. The UMTS air-interface is shown as a mobile S- interface (SM) [2].

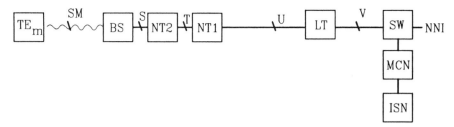

Figure 3 : Example UMTS Functional Architecture

Generic Functional Model

As a basis for the service specific functional models of the CCITT three stage methodology for signalling protocol development, a **generic** functional model of UMTS has been defined. Figure 4 depicts one half of the model, which is symmetric, and also indicates a possible allocation of functional entities to different parts of the network and r1-r16 correspond to relationships between various functional entities.

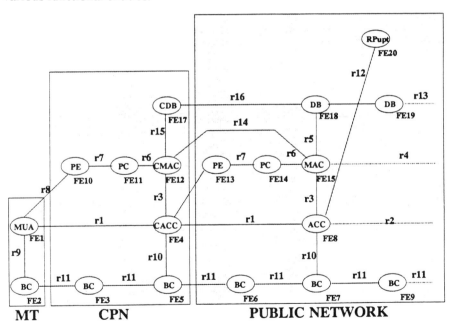

Figure 4 : UMTS Generic Functional Model

The generic functional model is in fact a model for call set-up, based on the assumption that a set of call handling functions are common to a number of teleservices to be supported. A fundamental point behind having a generic functional model of UMTS is the anticipated integration of UMTS into the broadband telecommunications networks, and thus the UMTS generic functional model is based on the generic B-ISDN functional model, enhanced with mobile system specific functions.

From the Functional Entity (FE) point of view, three layers exist; the functions in each layer are logically separated from the functions of the other layers:

- Mobility Call Control FE layer : Paging Entity (PE), Paging Control (PC), (Customer) Mobile Access Control (CMAC), , Data Base (DB), UPT Routing Point (RP$_{UPT}$)

- Call Control FE layer : Mobile User Agent (MUA), (Customer) Access and Call Control (CACC)

- Bearer Control FE layer : Bearer Control (BC)

Support of UPT has been incorporated into the functional model via the functional entity RP$_{UPT}$ showing the functions of the UPT service.

Mobility Issues

The following issues have been studied as a part of the mobility and data management in UMTS:

- Attachment/ detachment procedures as a means of informing the network of a terminal's (in)active status to prevent paging to inactive terminals.

- Locating and paging, which are performed at call set-up to locate a user on a mobile terminal. Because the scale of UMTS does not allow global paging, the area covered by the system is divided in smaller location and paging areas. Locating involves interrogation of a database that allows to find the current location area of a subscriber.

- Location registration, to keep location data in the network up-to-date as a user on a mobile terminal roams.

- Handover, to ensure continuity of calls when mobile terminals cross the boundaries of areas covered by different base stations.

- Security, in particular ciphering for confidentiality of information, and authentication for mutual identification of network entities.

Dominant mobility procedures such as location registration, locating, paging and handover have been described in terms of functional models and information flows.

Data Model

A data model has been proposed for UMTS, specifying the data necessary to operate UMTS and its distribution over the network. The starting point for a UMTS data model is the numbering and identification scheme. For this purpose of numbering and identification three kinds of users are distinguished in UMTS :

- Users who use only terminal mobility (in this case there is no distinction between the identification of user and terminal),
- Users who use terminal mobility and personal mobility within the boundaries of UMTS (UMTS User Mobility or UUM users),
- UPT (Universal Personal Telecommunications) users.

Figure 5 shows an overview of the UMTS data model in the form of an entity-relationship diagram.

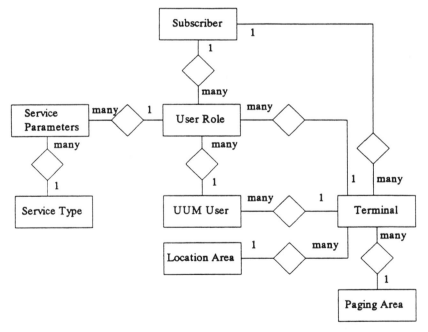

Figure 5 : UMTS Data Entities and their relations

Security

UMTS should offer at least a level of security comparable to that of fixed networks, despite the sensitivity of the radio link. Important security issues in UMTS are

- the authentication of users and network entities and,

- protection of user and signalling information on the radio link.

Authentication will be required in call set-up, location registration and handover. Authentication and key management will be based on CCITT Recommendation X.509.

Encryption is used to protect user and signalling information on the radio link. A public key algorithm is preferred for UMTS, considering technical advantages and taking into account trends in related fixed networks. However at the present state of technology, a public key algorithm can not be used to encrypt information at the bit-rates envisaged for UMTS.

A hybrid solution is proposed where initial signalling is encrypted with a public key algorithm, and after secret keys have been exchanged, a faster secret key algorithm takes over.

5. UMTS RADIO INTERFACE RESEARCH

Spectrum Requirement for UMTS

A requirement on the use of spectrum is that it should be possible to split the spectrum allocation in any area among several operators. The spectrum requirement for UMTS is based on the allocation of spectrum to third generation systems by the World Administrative Radio Conference (WARC) in 1992.

During WARC'92, a total of 230 MHz, in frequency bands 1885-2025 MHz and 2110-2200 MHz, was targeted, on the world-wide basis, for FPLMTS (UMTS being studied in ETSI SMG5, can be thought of a projected European standard of the FPLMTS).

In all ITU regions, 2x30 MHz (in region 2-U.S.A.-, this allocation was extended to 2x40 MHz plus some extra 'secondary' allocated spectrum) was 'co-primary' allocated for mobile satellites in the bands 1980-2010 MHz (uplink) and 2170-2200 MHz (downlink).

Radio Link Control Functions

The need for a dense micro-cellular and pico-cellular network in a multi-environment involving a variety of CPN's required special considerations in signalling design for functions like handover. These issues have been addressed by both Channel Management and Fixed Network study areas in the project.

By considering various combinations of possible cases, a total of 26 handover types have been identified. Following cases are to be considered :

- Handovers in Public/Domestic/Business and Mobile Environments,
- Intra/inter cell handovers,
- Inter BSC (Base Station Controller)/LE (Local Exchange)/MCN handovers,
- Single/multiple user (i.e.,for mobile CPN's handovers),
- Intra/inter environment handovers.

In micro-cellular networks where the BTS's (Base Transceiver Station) may be connected in ring topology, handover processing might be shared between adjacent BTS's. However, a final handover execution by the BTS is possible only if it can handle the radio resource allocation function. Ideas of handover from micro-cell to an umbrella macro-cell in case of inter-environment handover have been developed to some extent [2,6-7].

Handover studies assume Fixed Channel Allocation (FCA) schemes and circuit-switched connections in a Time Division Multiple Access (TDMA) based system. Cases of Mobile Assisted Handover (MAHO) and Mobile Controlled Handover (MCHO) are considered. Handover initialisation involves the usual set of radio link parameters like Received Signal Strength Indication (RSSI), Bit Error Rate (BER), interference measurements and the Mobile station distance from the cell sites. Considerations of correlating parameters measured at MS and BS are expressed.

However, it is suggested that the interference situations (especially for co-channel interference) at cell site and a particular MS are rather different. Moreover, it is found that different thresholds for handover initiation are required in different environments. Distinction between micro-cell and macro-cell networks, and between line of sight (LOS) and no line of sight (NLOS) situations are important. Some new handover initiation criteria like mobile speed, trend analysis and traffic load analysis are mentioned. No details concerning their calculation or application are available.

Moreover, no clear cut algorithms and handover performance evaluation figures are available. Dynamic Channel Allocation (DCA) is assimilated to intra-cell handover. It is suggested that mobile controlled DCA is not the most reliable algorithm for micro-cellular environments. No explicit studies on Adaptive Power Control (APC) are available and it has been suggested that both DCA and APC use the same physical resource. Issues like re-establishment of best connection path in the network, recovery from unsuccessful handover and the required Grade of Service (GOS) definition require further attention in the project [8].

Use of high traffic density micro-cellular networks reduces the size of manageable location areas such that the location updating signalling traffic increases too significantly. A multi-environment location updating requires specific checking for user integrity and service profiles. Studies show that location area size has much influence on signalling between base stations. It also affects the rate of failing paging attempts.

Certain specific considerations like peak subscriber density, high speed vehicles in small location areas and mobile CPN's require dynamic location area dimensioning. However, such a solution has been considered too complex for initial consideration, Cases of paging areas identical to, bigger than and smaller than location areas have been analysed and only paging areas identical or smaller than location updating areas are considered for further evaluation.

Multiple Access Techniques and Radio Signal Design

This section covers the research done for the design and evaluation of multiple access techniques and on the evaluation of modulation and equalisation schemes judged useful for UMTS applications.

Multiple access techniques in synchronised radio interface (TDMA) and packet transmission have been studied in detail [2]. Two carrier Time Division Duplex (TDD) with associated TDMA has received a lot of attention and a detailed study on DCA in this environment has been performed. Different DCA algorithms are:

- fully autonomous DCA controlled entirely by the mobile,
- regulated DCA where parameters are supplied by the base station,
- segregated DCA where the preferential channels are given by the base stations,
- combination of the two above.

Channel selection is based on RSSI and BER calculations. It has been found that fully autonomous DCA gives smaller capacity gain than the other algorithms and is considered unsuitable for micro-cellular networks due to rapid traffic fluctuations. DCA based on interference measurements performs better when associated with Slow Frequency Hopping (SFH) but this needs hopping, synchronisation across the network.

These studies have been performed essentially for bit-rates associated with speech transmission only and it is suggested that a range of DCA algorithms suited to different requirements should be available.

A separate work package considered the packet transmission in UMTS. Pure and Slotted Random Multiple Access and Packet Reservation Multiple Access (PRMA) protocols have been studied. PRMA is better suited for voice traffic than Time Multiplexed-Base Control Multiple Access with collision Detection (TM-BCMA/CD). This latter can cope better with variable bit-rate data traffic than the PRMA itself. PRMA is quite flexible and can operate as classical

TDMA or as a completely random access protocol. It has been concluded that packet transmission offers considerable performance improvement over TDMA in high bit-rate speech transmission in small cell sizes and also for bursty data transmission. However, the current tendency in the project is that the voice should be circuit-switched rather than packet.

Code Division Multiple Access (CDMA) is currently a fashionable radio access scheme and has been suggested as an alternative access method for the third generation system [5]. However, CDMA has not been studied to the required extent in the previous years during this project. Only recently some work indicating the relative advantages of CDMA have been identified [2,9]. For example, there is no need for frequency planning methods in a CDMA system. A CDMA system inherently makes use of interferer diversity, but frequency hopped TDMA gives similar advantages. In addition, there is some concern of CDMA's potential to support high bit-rate services and the possible high network infrastructure cost to support soft handover, in particular when UMTS is integrated into the IBCN.

Today, in Europe, there is no clear cut idea on which whether CDMA or TDMA is to be used for UMTS. Hopefully, RACE2 will answer this question in three years time.

The propagation work to characterise indoor pico-cells and outdoor micro-cells at 1.7 GHz has been performed by Radio Bearer and Cellular Coverage study groups in the project. For outdoor macro-cells use of duly modified COST-207 models has been suggested. One again, the radio channel characterisation for narrow band TDMA channels is considered. In addition to data on channel time dispersion and signal strength profiles, some measurements of man-made noise at 1.7 GHz are the results of this work. However, a clear cut channel model for indoor pico-cells only has been proposed and used for further work on radio signal design.

The signal design work involving modulation and equalisation studies was handled again by RB-MEC study groups. Studies on Linear equalisers, Decision Feedback equalisers, hybrid schemes like Decision Feedback Sequence Estimators (DFSE) and reduced state Viterbi type of equalisers have revealed that it is possible to reach data rates of 8 Mb/s in indoors and 1 Mb/s for outdoor macro-cell channels [2]. Also the results showed that it is possible to achieve such data rates with a tolerable receiver complexity and a BER of 10^{-2} valid for low bit-rate speech transmission in a co-channel interference limited environment. These evaluations were performed with simulations using Common Channel Interface (CCI) developed by a sub-group in RB core area. CCI is useful tool for only narrow band TDMA channels, in which GSM type of burst structures have been used.

Different values of channel bit-rates varying from 300 kb/s to 8 Mb/s have been simulated with short and long TDMA bursts. As the outdoor channel models are translated version of COST-207 models, the effect of higher order Doppler components on the performance of different equalisers have not been evaluated. Attempts to compare the implementation complexity of these equalisers have also been made at various occasions in the project. DFE in general offers lower complexity at the expense of lower performance. A fixed burst structure with a training sequence at the centre is proposed.

As far as modulation scheme is concerned, various modulation techniques like constant envelope (GMSK, GTFM which is Gaussian Tamed Frequency Modulation) and linear modulation (QAM, bandlimited QPSK) are studied for UMTS design. In choice of modulation schemes, the spectrum efficiency (in Erlangs/km2/MHz) performance against co-channel and adjacent channel interference, support for an adaptive bit-rate, need of an equaliser and complexity play an important role. Linear modulation schemes such as QAM and QPSK are generally more spectral efficient than constant envelope ones, but require the use of high efficiency linear power amplifiers.

The above equaliser schemes have been tested with GMSK and 4-QAM modulation schemes. As the variable bit-rate transmission is receiving a lot of attention in the project, multi-level modulations of both constant and non-constant envelope types are to be studied further. Another possibility is to use Trellis Coded Modulation (TCM) for variable bit-rate transmission and that has recently received a lot of attention. Results from equaliser studies to deduce peak power requirements for various cell types are given in Table 1.

Environment	Bit-rate	Range	Peak Power
Indoors	2.0-8.0 Mb/s	20-30 m	100 mW
Urban	500 kb/s	1.0 km	1 W
Hilly Terrain	300 kb/s	tbd	tbd
Rural	300 kb/s	6.0 km	1 W
Street Microcell	2.0 Mb/s	200-300 m	100 mW

Table 1 : Power Budget Calculations

Both CM and MEC groups have undertaken studies on error correction mechanisms. CM concentrated more on ARQ type techniques and MEC on use of FEC methods. FEC methods have involved the use of Reed-Solomon, BCH codes and 1/2 rate convolution codes. These coding schemes have been applied to the output files from equaliser studies. One of the main result is related to the employment of hybrid FEC/ARQ using a BCH at less than full power as it produces a reasonable throughput even if an error rate of 10^{-8} is asked. CM studies indicate that FEC schemes tend to degrade throughput and require long interleaving which increases transmission delay. On the other hand ARQ schemes are simple and more flexible to implement but, suffer from variable delay.

Speech Coding

Since the UMTS aims at the dispatch of a large amount of traffic, it requires efficient techniques to reduce the bit-rate of connected sources. In selection of speech coding, many conflicting factors such as bit-rate and spectrum efficiency, quality, robustness to transmission errors, delay and power consumption, cost and complexity are to be carefully considered. The UMTS speech codec attempts to meet the constraints given by the CCITT ad-hoc group for a third generation system.

These are:

Speech quality :	>	G721.Rec.	
One way algorithmic delay :	<	10	ms
Source bit-rate :	<	8	kb/s
Gross bit-rate :	<	16	kb/s
Complexity :	<	GSM	

In this project, all the speech coding activities were carried out by Signal Processing core area during the first three years of the project. Four different coding algorithms have been designed and evaluated [2]. Two of these are improved version of GSM full rate codec and use Regular Pulse Excited (RPE) linear prediction techniques. They are referred to as RPE-HLTP (RPE with High Resolution Long Term Prediction) and LDP-RPE (Low Delay Prediction with RPE). The other two rely more on methods using vector quantisation (VQ) and they are referred to as LAS-VQ (Long Term Predictive ADPCM with Short Term Prediction and VQ) and SVE (Singular Value Excitation Coder).

Laboratory evaluation of these codecs with clean speech and speech with ambient disturbances have been done. Only the codecs with long algorithmic delay of 20 ms were able to offer good MOS (Mean Opinion Score) performance, but the performance degrades rather steeply when they are put in tandem or are operating in presence of ambient disturbances. VQ based codecs through of high complexity of implementation are more robust to these disturbances and require lower algorithm delay as well.

Table 2 summarises the source bit-rate without error protection, delay, complexity (GSM codec is selected as a bench-mark), and quality in MOS for clean speech. Today, no codec is known to be good enough. The characteristics of codecs for GSM (RPE-LTP), D-AMPS (VSELP), GSM-HR (half rate GSM codec [9]) and DECT (ADPCM) are also included for an interest [10-11].

Codec Type	Bit-Rate in kb/s	Delay in ms	Complexity wrt GSM	MOS
LAS-VQ	8.10	7.50	5.0 * GSM	3.5
RPE-HLTP	7.90	20.0	2.0 * GSM	3.8
SVE	8.30	8.0	5.0 * GSM	2.9
LDP-RPE	13.2	5.0	1.5 * GSM	3.2
GSM Codec	13.0	20.0	1.0	3.8
D-AMPS Codec	8.0	20.0	3.0 * GSM	3.4
GSM-HR Codec	6.5	20.0	4.0 * GSM	3.5
DECT Codec	32.0	0.125	0.5 * GSM	4.5

Table 2 : Characteristics of Speech Codecs

It seems that a speech codec satisfying all the CCITT requirements will be much more complex than the GSM full rate codec and will require very elaborate error correcting mechanisms

RACE-Mobile Systems Group concludes that a candidate codec should achieve a high quality with limited error protection under the frame loss rates which are expected over the fast fading mobile channels. Further, the performance of this low bit-rate codec at a reasonable cost could be improved with variable bit-rate techniques.

6. SYSTEM LEVEL WORK FOR UMTS

A Systems Group (SG) was established to complement the bottom-up research approach of the individual study areas with top-down system design. This group basically focussed on four major working documents: UMTS User Requirements (extension of UMTS Requirements Specification [3] given to ETSI), Introduction and Evolution of UMTS, System Description Document (SDD) and System Framework (SF). The outputs of the SG are intended to provide standardisation and other bodies with the salient results of RACE Mobile and a framework on which to build when specifying the third generation system. Many of the issues raised in these documents are summarised in the Common Functional Specifications (CFS) produced by RACE Industrial Consortium (RIC). In particular, the CFS most relevant to mobile is the CFS/D730 [2]. This document has been presented to CCIR TG8/1.

SDD is the core of the RACE-Mobile project and summarises all the technical work done in overall project and then gives conclusions and serves as a reference to other deliverables in the project.

Introduction and Evolution of UMTS document discusses the introduction and development of the UMTS against the future trends in telecommunications at the turn of century (such as GSM/DCS1800, CT2, DECT, N-ISDN and UPT). Since UPT will be available on N-ISDN, PSTN, GSM (end 1995) and other networks by the late 1990's well before the introduction of UMTS, this document discusses the inter-network personal mobility offered by UPT service and sees as complementary to UMTS. It concludes that as UPT will facilitate access to UMTS from a user's point of view, UPT and UMTS will benefit from each other.

User Requirements document is the extension of UMTS Requirements Specification input [3] given to ETSI. The latter outlines the basic requirements assumed by RACE Mobile, as seen from the users and network operators point of view. The former one includes some details of services (stage I descriptions), environment, handover-roaming requirements, mobility, terminals, service usage characteristics/markets and network management.

Figure 6 depicts some load estimates for environments based on the assumed parameters and usage characteristics of 5 example services : telephony (8kb/s), telefax (64 kb/s), video telephony (64 kb/s), database retrival (64 kb/s) and a high bit-rate service (500 kb/s).

The SG activities have been centred around the development of a system framework. The system framework results from research work undertaken in the project in specialised technical core areas, and combines the individual results and makes conclusions towards a UMTS system design. Primarily, it concentrates on user requirements, the adaptivity requirements and resulting in

48

detail of the lower layers of the radio interface, and considers some deployment issues with reference to an "example town". This work intends to be a paper "**test-bed**" forming a strong basis for the coming work in RACE2 studies. The methodology and key results of the SF are decsribed in [12].

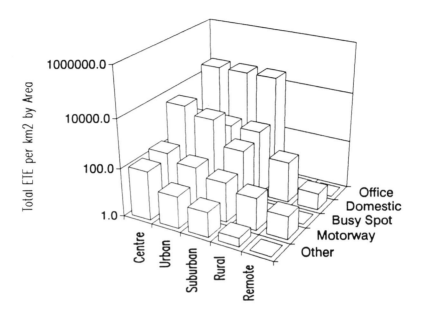

Figure 6 : Traffic Load Estimates per km² per area type

For this exercise, a sub-set of the UMTS user requirements have been taken and interpreted into both quantitative and qualitative assumptions governing the system detail design. Table 3 shows the current system choices and relative parameters for an air-interface of a UMTS (these are only examples). The choice of the multiple access method purely reflects the fact that most of the RACE Mobile work focussed on TDMA access techniques.

The need of adaptive bit-rate is well recognised in the project and therefore two carrier bit-rates have been selected based on the modulation, equalisation and propagation studies performed in the project. In order to support high bit-rate services, common carrier bit-rate for pico and micro-cells has been chosen. Notice that macro-cell bit-rate is significantly lower due to delay spread, speech delay and transmitting power restrictions. The use of multi-level modulation with a selectable number of levels in pico-cells emphasises the adaptivity in bit-rate requirement.

Air-interface Parameters	Pico-cells	Micro-cells	Macro-cells
Cell radius	< 100 m	< 1 km	< 35 km
Multiple Access	TDMA/FDMA		
Carrier Bit-rates	2.9 Mb/s		484 kb/s
Modulation Scheme	QAM		
Burst Structure	Voice : 44 data + 32 overhead Other : 120 data + 32 overhead		
Duplexing	TDD	FDD	
Slots per Carrier	96		16
Access Networks	B-ISDN, ISDN and PSTN	B-ISDN and ISDN	
Logical Channel Allocation	Circuit and Packet		Circuit
Speech Coding Rate	8 and 32 kb/s		8 kb/s
Handover Control	Mobile Controlled	Network Controlled with call re-establishment if required	

Table 3 : Some initial UMTS System Framework parameters for the air-interface

For pico-cells, TDD was selected as it offers greater flexibility for frequency allocation in the less regulated environment, while FDD was chosen in macro-cells to reduce peak transmit power and to avoid TDD delay overheads. FDD was also selected for micro-cells as it allows simpler handover. Handover between micro-cells and macro-cells is likely to be a frequent event with macro-cells being used as umbrella coverage over non-continuous micro-cell regions. On the other hand, handover between pico-cells and other types is expected to be less frequent and, hence a short interruption of service could possibly be tolerated.

Circuit switched channel allocation is supported to facilitate simple access. Packet access is also supported in micro and pico-cells to provide more spectrum efficiency for bursty services.

Network controlled handover leads to a more spectrum efficient system in micro and macro-cells. Call re-establishment can also be used when network control fails due to unexpected propagation conditions. In pico-cells, mobile controlled handover seems to be preferable (environment where spectrum efficiency is less critical) as it provides a quicker response to varying propagation conditions.

The system framework evaluation was based on a example town model, characterised by a population profile and usage characteristics. Presently, the parameters of the model are derived for the city of Amsterdam.

Although, the work requires more studies, it seems that a UMTS design to meet the requirements is feasible. Micro-cells play an important role, especially economic feasibility of the system highly depends on the micro-cell infrastructure. Hence, any approach to minimise this cost must be one of the main objective in designing third generation system like UMTS.

7. MISCELLANEOUS ISSUES

This chapter is devoted to the work done in three specific areas such as techno-economics, implementation and Mobile Broadband Systems [2]. The following sub-sections try to summarise these results.

Techno-ecomonics of the UMTS

With the total 230 MHz allocated by WARC'92, micro-cells will be essential to support the traffic expected in UMTS. A range of cells will, however, be necessary due to infrastructure cost considerations. Some preliminary results in the project give the total number of macro-cells and micro-cells for a typical area. It should be emphasised that these are heavily dependent on the traffic assumptions and models used. For example, with 200 MHz spectrum allocation, in order to cover an area of size of England and Wales, one may need 4000 macro-cell base stations and around 24000 micro-cell base stations.

These figures as stressed are highly dependent on traffic characteristics. In addition, the infrastructure cost was worked out, the main part of the costs involved lies in the base station infrastructure. Therefore, it was concluded that, from the techno-economics point of view, the techniques and technical solutions (like network topology -star or ring- may have to be different for different radio network types-city micro-cells, motorway micro-cells and rural macro-cells) which keep the costs down should be found.

Implementation Issues

The project also looked at the hardware components and evaluated the impact of various parameters upon the handset, base stations and other network elements in terms of size, complexity, power consumption, performance and cost. The expected developments in technology of ten years from today also considered, with particular emphasis on the low-cost pocket handset.

Handsets consist of many parts such as keypad, antenna, battery, receiver front-end (mixers, AGC amplifiers, filters), power amplifier for transmitter, DSP (speech coding/decoding, channel coding/decoding, equaliser, channel filtering). Both cost and the power consumption figures for various stages of handset based on 1990s technology were estimated as a best guess for the year 2000. This reveals that handset cost for UMTS is around 100 ECU ($120) for a mass use. In addition, some work has been reported on the base station, network elements implementations and other components.

MBS Contribution to RACE Mobile Project

RACE Mobile project also includes the Mobile Broadband System, studying the provision of broadband bit-rate services, potentially between 1 to 155 Mb/s such as broadband quality mobile video. This specific work in the project looked at a two distinct aspects : transmission at 60 GHz in mobile environment and very wideband transmission for multi-media communications. After an analysis of traffic requirements for multi-media services at the turn of century, a traffic model was established. In parallel, propagation studies for wideband transmission in micro-cellular and pico-cellular environments were conducted. Some technological developments related to semi-conductor device design have also been reported. Most of this work was done during phase I of RACE1. However, both UMTS requirements specification given to ETSI and the UMTS system framework do not include any considerations for broadband system design in RACE1.

8. CONCLUSIONS

In this paper, an attempt has been made to report on the progress of technical work done in various study areas by RACE Mobile project. As can be seen, this project has concentrated on the use of TDMA techniques to support multi-environment mobile communication systems. A detailed study of certain aspects like radio signal design and fixed network issues performed during phase I of RACE1 has provided very useful input for system framework definition during phase II of RACE1. The system framework, which is based on TDMA, is the major document in this project that combined the individual work done in specialised technical areas and made conclusions towards a UMTS as a third generation system.

Due to world-wide interest in CDMA for mobile communications, RACE Mobile project is also examining more and more issues related to this subject.

The RACE Mobile project will be finishing later in 1992. It will be replaced by a set of RACE2 projects which will individually consider issues including mobility in the network, frequency planning tools, CDMA and TDMA radio access and mobile broadband. These new projects will take the existing work on to a more practical phase with the development of test beds, demonstrators and will also prepare inputs to standardisation bodies.

ACKNOWLEDGMENTS

The RACE Mobile project consortium consists of many European Manufacturers, Operators and Research Institutions. Philips Radio Comm. Systems (UK), AEG (D), Alcatel Radiotelephone (ART-F), Bosch (D), BBC (UK), British Telecom.(UK), Thomson Hybrids (F), RNL Research Labs.(NL), EB Tech.(N), Ericsson Radio Systems (S), Fondazione Ugo Bordoni (I), GEC-Marconi (UK), Mullard (UK), National Microelectronics RC (IRL), Nokia Corporation (SF), Hellenic Telecom. Organisation (GR), Philips Research Labs. (PRL-UK), Philips Komunikations Industrie (PKI-D), Telecommunications Radioelectriques et Telephoniques (TRT-F), Alcatel SESA (E), Telettra (E), Thorn EMI (UK), University of Strathclyde (UK), Roke Manor Research (Siemens-UK) are the partners of this consortium.

The authors would like to thank all the members of the RACE Mobile Consortium and in particular, A. Dennis (PRL), C. Loison (ART), A. Urie (ART), H. Zuidweg (RNL) and the other members of the RACE Mobile System Group.

It is important to note that any opinions or judgements presented in this paper should not be taken as the consensus view of the entire project but treated as the personal view of the authors.

REFERENCES

[1] **D.Grillo and G.MacNamee**, "European Perspectives on third generation Personal Communication Systems", IEEE-VTC Conference Proceedings, Orlando, U.S.A., May 1990.

[2] **Race Industrial Consortium**, "Mobile Network Sub-systems", RACE Common Functional Specifications, CFS/D730, Issue B, January 1992.

[3] **Race Industrial Consortium**, "RACE-UMTS Requirements Specs.", Doc. no.16/91, ETSI-SMG5, Oslo, Norway, November 1991.

[4] **L.M.Sellar**, "The RACE Mobile Project", MRC Proceedings, Nice, France, November 1991.

[5] **A.Salmasi and K.S.Gilhousen**, "On the System Design Aspects of CDMA applied to Digital Cellular and Personal Communications", IEEE-VTC Conference Proceedings, St. Louis, U.S.A., May 1991.

[6] **J.Grond, H.Hecker and A.Wilhelmus**, "Future Generation in Mobile Communications", ITU Telecom. Conference Proceedings, Geneva, Switzerland, October 1991.

[7] **H. de Boer, M.Meijer and E.Buitenwerf**, "Network Aspects for the third generation mobiles", IEEE-Globecom. Conference Proceedings. Phoenix, U.S.A., December 1991.

[8] **CCITT : E750** Series of Recommendations- " Traffic Engineering Aspect of Mobile Networks ", SGII Plenary Meeting, Geneva, Switzerland, February 1992.

[9] **A.Baier and W.Koch**, "Potential of CDMA for third generation mobile systems", MRC Proceedings, Nice, France, November 1991.

[10] **C.C.Evci**, "Speech Codec Aspects for third generation mobile systems", IEEE-VTC Conference Proceedings, Denver, U.S.A., May 1992.

[11] **R.Montagna and P.Usai**, "Selection test for GSM half-rate channel", MRC Proceedings, Nice, France, November 1991.

[12] **A.Dennis, M.Streeton and A.Urie**, "A system framework for third generation mobiles", IEEE-VTC Conference Proceedings, Denver, U.S.A., May 1992.

The Use of SS7 and GSM to Support High Density Personal Communications

Kathleen S. Meier-Hellstern[1] and Eduardo Alonso
Wireless Information Network Laboratory
Rutgers - The State University of New Jersey
P.O. Box 909
Piscataway, New Jersey 08855
Phone: (908) 932-5954   Fax: (908) 932-3693

Douglas R. O'Neil
BellSouth Enterprises
1100 Peachtree St., N.E.
Atlanta, Georgia 30309
Phone: (404) 249-4503   FAX: (404) 249-5451

Abstract

The combination of high user density, high mobility and enhanced network capabilities in Personal Communication Networks (PCN) will generate significant network signaling. This paper presents an organized method for quantifying the Signaling System Number 7 (SS7) traffic load associated with terminal mobility when the Pan-European standard GSM is used for PCN. We also quantify the relative switching cost of PCN so that switching and signaling costs can be traded off to optimize network configurations for different PCN networks. Using CCIR parameters for an extremely dense, outdoor center-city pedestrian environment we illustrate the possible heavy signaling load. For the example parameters, the additional SS7 network burden is 4 - 11 times greater for cellular than for ISDN and 3 - 4 times greater for PCN than for cellular, depending on the Visitor Location Register placement. Our numerical results are very sensitive to assumptions on user mobility, cell size and location area size. This highlights the need for careful investigation of these parameters in specific applications.

---

1. This work was completed while the author was a Visiting Scholar from AT&T Bell Laboratories.

1. Introduction

Personal Communication Networks (PCN) are an evolution of the current cellular concept to an environment characterized by high user density, high mobility and enhanced network capabilities. Users will carry lightweight, low power handsets, enabling them to make calls from anywhere at any time. Each user has a single "Personal Telephone Number", and network procedures route calls to the appropriate wireless network. Significant network signaling is required to support personal and terminal mobility in PCNs. Personal mobility allows a user to access services at any terminal in accordance with the user's service profile. The user can receive incoming calls at that terminal or direct them to some other location. Terminal mobility allows a terminal to access the network at different locations as a user moves about.

This paper studies terminal mobility in the Pan-European digital cellular standard GSM (originally Groupe Speciale Mobile, now Global System for Mobile Telecommunications) [1,4]. The signaling traffic for GSM PCN differs from cellular in the higher penetration, smaller cells (microcells vs. macrocells) and location areas, and potentially increased calling rates. Since the signaling procedures for most current and planned wireless networks are similar to GSM, the results have broader implications.

Based on predictions of user density and mobility [3,5], the additional signaling load could be substantial. Noting that personal mobility will add further to the network burden, this paper underlines the careful consideration that needs to be given to the modeling of mobile users as well as to signaling network planning.

2. GSM and the Intelligent Network

GSM has been chosen to provide PCN in Europe.[2] The GSM system conforms to the Intelligent Network concept (IN) [11], which uses Signaling System Number 7 (SS7) [14] for communication between network elements. The SS7 network is a highly reliable, out-of-band, packet switched network, interconnected using 56 kb/s (US) and 64 kb/s

---

2. A group was formed at ETSI (European Telecommunications Standards Institute) to review and modify the GSM recommendations to provide Personal Communication Services in the 1800 MHz band. The standard uses the same basic signaling procedures as GSM and will be known as DCS1800.

(Europe) links. In GSM, SS7 interconnects Mobile Switching Centers (MSC), Visitor Location Registers (VLR) and Home Location Registers (HLR) [1,4]. The MSC sets up calls to mobile users via the base stations and maintains connections to other MSCs and the Public Switched Telephone Network. The HLR stores the permanent subscriber parameters and features for a group of subscribers within a network. It contains pointers to VLR's to assist in routing incoming calls. The Visitor Location Register (VLR) is a local database in charge of one or more location areas. It contains the subscriber parameters of all subscribers currently within these areas. The VLR obtains subscriber parameters from the HLR, and updates the HLR regarding the status of special services, if necessary. It updates the HLR as terminals move into its area and performs authentication for mobile subscribers. The VLR may also allocate temporary, location-specific, phone numbers. The VLR and HLR databases (consistent with the Intelligent Network concept) can be located in or outside the MSC.

3. Signaling Network Load

Call setups, inter-MSC handovers and location updates are the key activities which generate SS7 traffic in GSM. Detailed message flows and message sizes for these activities may be found in [1,4,7]. SS7 message lengths are variable. Our estimates include only mandatory parameters and allow for the maximum parameter length of those parameters that are included. (For ISDN messages we also include several optional parameters).

Call setups are comprised of mobile originations (calling party procedures) and mobile terminations (called party procedures). For a mobile origination, the number of bytes between the SS7 network entities is:

— Originating MSC and terminating switch: 120 bytes,

— Originating MSC and associated VLR: 550 bytes.

For a mobile termination, the number of bytes between the SS7 network entities is:

— Originating switch and terminating MSC: 120 bytes,

— Terminating MSC and associated VLR: 612 bytes,

— Originating switch and HLR: 126 bytes.

58

Inter-MSC handovers transfer a call in progress to a new MSC. For an inter-MSC handover, the number of bytes between the SS7 network entities is:

— New MSC and associated VLR: 148 bytes,

— New MSC and old MSC: 383 bytes.

Location updates record the locations of subscribers as they move through the network. Whenever a user moves into a new location area (group of cells), a location update is required. A VLR controls one or more location areas, and slightly different procedures are used depending on whether the mobile moves into a new VLR area or stays within the existing VLR area. For a location update within the same VLR, the number of bytes between the SS7 network entities is:

— MSC and associated VLR: 406 bytes,

— VLR and HLR: 55 bytes.

For a location update to a new VLR, the number of bytes between the SS7 network entities is (Figure 1):

— New VLR and associated MSC: 406 bytes,

— New VLR and old VLR: 213 bytes,

— Old VLR and HLR: 95 bytes,

— New VLR and HLR: 182 bytes.

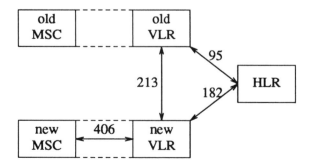

Figure 1. SS7 Location Update Traffic

Table 1 aggregates the message traffic by network element. We have assumed no mobile-to-mobile calls. The first row of Table 1 shows the number of bytes sent to and from the switch for a standard ISDN call

| | Orig. | Term. | Inter-MSC Handover | | Loc. Upd.- Same VLR | Loc. Upd.- New VLR | |
| | | | MSC old | MSC new | | VLR old | VLR new |
	b_1	b_2	b_3	b_4	b_5	b_6	b_7
to/from Switch: ISDN Call	120	120	-	-	-	-	-
to/from MSC: VLR in MSC	120	120	383	383	55	308	395
to/from MSC: VLR out MSC	670	732	383	531	406	-	406
to/from VLR	550	612	-	148	461	308	801
to/from HLR	-	126	-	-	55	95	182
Internal SS7: VLR in MSC	120	246	383		55	490	
Internal SS7: VLR out MSC	670	858	531		461	896	

TABLE 1. SS7 Bytes Generated per Transaction for Key Wireless Activities

with a minimum of optional parameters. The second and third rows show the number of bytes sent to and from the MSC assuming that the VLR is either inside or outside the MSC, respectively. In both cases, the HLR was external to the MSC. The fourth and fifth rows show the number of bytes sent to and from the HLR and VLR, assuming that they are both external to the MSC. The sixth and seventh rows show the number of bytes that appear internal to the SS7 network, depending on whether the VLR is inside or outside the MSC.

To see how the table entries are computed, Figure 1 displays SS7 traffic between network entities for location updates to a new VLR. The dashed lines between the MSCs and their associated VLRs indicate that the VLR can be inside or outside the MSC. Consider the last two rows of Table 1 which give the internal SS7 traffic. When the VLR is in the MSC, we see from Figure 1 that the number of internal SS7 bytes is 490 (213 + 95 + 182). When the VLR is outside the MSC, the number of bytes is 896 (406 + 213 + 95 + 182).

3.1 Frequency of Transactions

In order to aggregate the information in Table 1 into the total traffic offered to the SS7 network, we compute the frequency of originations, terminations, handovers and location updates per hour. The needed quantities (as a function of the MSC serving area) can be computed as follows.

- *MSorig(area)* and *MSterm(area)* are the number of mobile originations and terminations, in *calls/hour*, generated in *area*.

$$MSOrig(area) = area * \frac{people}{km^2} * P[person\ has\ terminal] \qquad (1)$$
$$* P[mobile\ origination] * \frac{erlangs}{terminal} * \frac{1}{call\ holding\ time} .$$

$\frac{people}{km^2}$ refers to the total density of people in the area, whether they are equipped with terminals or not. $\frac{erlangs}{terminal}$ refers to the number of erlangs generated per terminal, and *P[mobile origination]* is the probability that a call is mobile originated. The formula for mobile terminated calls is identical to that for mobile originated calls with *P[mobile origination]* replaced by *P[mobile termination]* = $1 - P[mobile\ origination]$.

- *Handovers(area)* is the total number of cell departures (intra- plus inter-switch) made by calls in progress in the switch area in *crossings/hour*.

$$Handovers(area) = ncells * \frac{crossings}{cell} \qquad (2)$$
$$* P[person\ has\ terminal] * \frac{erlangs}{terminal} .$$

$ncells = \frac{area}{cellarea}$ is the total number of cells in the area, and $\frac{crossings}{cell}$ is the number of people (with or without terminals and with or without a call in progress) that leave the cell per hour (see Section 3.2), in *crossings/hour*. $\frac{erlangs}{terminal}$ is the total number of erlangs per terminal, which also corresponds to the probability that a call is in progress.

The number of inter-switch handovers generated by leaving the switch area, which is also the number of handovers into the area, is given by

$$InterMSCHandovers(area) = \frac{crossings}{area}$$
$$* P[person\ has\ terminal] * \frac{erlangs}{terminal}\ , \qquad (3)$$

where $\frac{crossings}{area}$ is the number of people that leave the area per hour.

The number of intra-MSC handovers is given by

$$IntraMSCHandovers(area) = \qquad\qquad\qquad (4)$$
$$Handovers(area) - InterMSCHandovers(area)\ .$$

Since for intra-MSC handovers, all departures from one cell are arrivals to a new cell in the same switch, only the cell departures need to be counted. Also, departures from the MSC area need to be excluded.

In an actual GSM system, groups of cells are controlled by a Base Station System (BSS). Intra-BSS handovers may be controlled solely by the BSS, depending on the system capabilities. In this case, (4) may be partitioned into intra- and inter-BSS handovers.

- *LocUpdates(area)* is the total number of location updates (intra- plus inter-VLR) in the switch serving area, in *crossings/hour*.

$$LocUpdates(area) = nlocareas * \frac{crossings}{locarea} * P[person\ has\ terminal]$$
$$* (P[terminal\ powered\ on] - \frac{erlangs}{terminal})\ , \qquad (5)$$

where *nlocareas* is the number of location areas in the switch coverage area, given by *nlocareas* $= \frac{ncells}{cells/locarea}$. $\frac{crossings}{locarea}$ is the number of people (with or without terminals and with or without a call in progress) that leave the location area per hour (see Section 3.2), in *crossings/hour*. The term $P[terminal\ powered\ on] - \frac{erlangs}{terminal}$ arises from the fact that only terminals that are powered on and do not have a call in progress generate location updates. (The location update is performed at the end of the call in this case. This is not included here, but will have a minor effect for the parameters of interest). The quantities *LocUpdSameVLR(area)* and

LocUpdNewVLR(area) can be determined in in the same manner as intra- and inter-MSC handovers.

3.2 Mobility Model

In order to calculate equations (2) - (5), a mobility model is needed to compute the number of cell, location area and switch area crossings. We use a simple flow-based model described in [9,13]. The model assumes that people are uniformly distributed in an area A, and that the direction of travel of each user relative to the border is uniformly distributed on $[0,2\pi)$. If we define ρ to be the density of people per km^2 (with or without terminals), v to be a person's average speed in km/hr, and L to be the length of the perimeter of the area A, then the average number of people leaving the area A per hour is given by

$$\frac{\rho vL}{\pi} . \qquad (6)$$

By conservation of flow, (6) is also the number of crossings into the area. In applying (6), it is important to relate ρ and v [5]. (This is emphasized in [12], where more general mobility models are considered.) Also, expressions are needed for L as a function of A for particular cell geometries. These are derived in [8].

3.3 SS7 Traffic per Call

Equations (1) - (5) can now be combined with Table 1 to compute the number of SS7 traffic bytes per call as a function of the switch serving area:

$$
\begin{aligned}
SS7\ bytes/call(area) = [&b_1*MSOrig(area) + b_2*MSTerm(area) \\
&+ (b_3+b_4)\ *InterMSCHandovers(area) \\
&+ b_5*LocUpdSameVLR(area) \qquad (7) \\
&+ (b_6+b_7)*LocUpdNewVLR(area)] \\
&/[MSOrig(area) + MSTerm(area)]\ .
\end{aligned}
$$

Remarks:

— The quantities b_i are the bytes per transaction taken from the relevant row of Table 1.

— The label "SS7 bytes/call" may be somewhat confusing, since location area update traffic, which is not call related, is included in (7). The per call normalization makes comparison with non-wireless signaling traffic easier.

— (7) is not a linear function of the switch serving area since (1) and (2) are proportional to area, but (3) is not.

— Only mean values are used in (7). Additional capacity is needed for variations above the mean.

— We implicitly assume that the number of mobile-to-mobile calls is negligible. If desired, these could easily be incorporated by adding an additional term in (7).

3.4 Switching Cost of a Wireless Call

It is tempting to conclude from (7) and Table 1 that the VLR should be placed inside the MSC to minimize SS7 traffic. However, when databases are placed inside the switch, the MSC must devote resources which could otherwise be used for call processing, to database transactions. This cost must be traded off against the SS7 savings in order to determine the appropriate network configuration. The switching cost of a wireless call can be computed in a similar manner to the signaling cost computed in (7), by replacing the weights b_i with weights that reflect the switch processing time and bandwidth consumed relative to wireline calls. Let these weights be denoted by w_i. For example, w_1 is the relative cost of a mobile origination compared to a wired origination. Since mobile-originated calls require interaction with the VLR for retrieving the call record and for authentication, $w_1 > 1$.

Equations (8) and (9) give the relative switching cost of a wireless call compared to a wired call. The equivalent calls per hour measures the load relative to wired calls generated by all of the activities (origination, terminations, handovers and location updates) associated with wireless calls. Using the weights w_i, this load can be expressed as:

$$
\begin{aligned}
Equivalent\ Calls/hour(area) = {} & w_1*MSOrig(area) \qquad (8)\\
& + w_2*MSTerm(area)\\
& + w_3*IntraMSCHandovers(area)\\
& + w_4*InterMSCHandovers(area)\\
& + w_5*LocUpdSameVLR(area)\\
& + w_6*LocUpdNewVLR(area)\ .
\end{aligned}
$$

For a given switch serving area, the equivalent calls per hour can be normalized, assuming the number of originations and terminations in the area is the same whether or not the calls are wireless.

$$Relative\ switching\ cost\ of\ wireless\ call(area) = \qquad\qquad (9)$$

$$\frac{Equivalent\ Calls/hour(area)}{MSOrig(area)+MSTerm(area)} \cdot$$

Remarks:

— (9) does not refer to the dollar cost of the switch, but rather indicates how the switching capacity is affected.

— (9) differs from (7) by including intra-MSC handovers, which use switch resources but not SS7 resources.

— MSC capacities are often quoted in calls/hour without regard to the underlying mobility assumptions. However, (8) is extremely sensitive to mobility assumptions. By defining the equivalent load, (8) can be used to account for varying assumptions on user mobility, and to give a more accurate representation of switch capacity.

4. Examples

The model from Section 3 can be used to compare GSM PCN, GSM cellular and ISDN traffic. As an example, we use CCIR estimates [3] of PCN traffic in an extremely dense outdoor center-city pedestrian environment. We note that the parameters in [3] appear to be quite high. In addition, mobility parameters which were not supplied in [3] had to be estimated. We estimated the pedsestrian velocity from an empirical model for "student pedestrians" given in [5].

We select cellular parameters assuming the same underlying environment with a lower penetration of cellular terminals and a smaller probability that a terminal is powered on. Table 2 displays the parameters.

Figure 2 plots internal SS7 bytes per call, using (7) and weights b_i from the last two rows of Table 1, versus MSC calls per hour. (Note that calls per hour is directly related to the switch serving area). The figure compares PCN, cellular and ISDN. For PCN and cellular, observe that the number of SS7 bytes per call decreases as the switch size increases. This is due to the fact that the number of inter-MSC location updates decreases as the switch serving area increases. The same is true of the number of inter-MSC handovers, but the effect is less pronounced since the volume of these is much less. The effect is non-linear since the mobility model defined in Section 3.2 depends on perimeter rather than area.

Parameter	PCN	Cellular
$\dfrac{people}{km^2}$	45,000	45,000
v	5.7 km/hr	5.7
$P[person\ has\ terminal]$	.8	.1
$P[terminal\ powered\ on]$	1	.5
$\dfrac{erlangs}{terminal}$	.04	.04
$P[mobile\ originated\ call]$	.5	.5
call holding time	2 min	2 min
length of cell side (assuming square cells)	.15 km	1.0 km
cells per location area [8]	39	13
VLR area	MSC area	MSC area

TABLE 2. Cellular and PCN Traffic Parameters

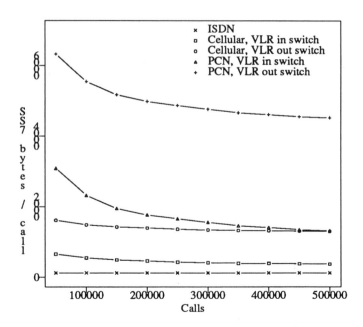

Figure 2. SS7 Costs for ISDN, Cellular and PCN

As an example, consider an MSC capacity of 200,000 busy hour call attempts. The number of SS7 bytes per call is 120 for ISDN, 459 for

66

cellular with the VLR in the switch, 1394 for cellular with the VLR out of the switch, 1768 for PCN with the VLR in the switch, and 4985 for PCN with the VLR out of the switch. Thus, the SS7 network burden increases by a factor of 4 - 11 for cellular compared to ISDN, and by a factor of 3 - 4 for PCN compared to cellular, depending on whether the VLR and HLR are located inside or outside of the MSC. For this example, the additional SS7 burden is substantial.

Figure 3 quantifies the relative switching cost of the different cellular and PCN options. When the VLR is inside the MSC, we chose the weights w_i to be $w_1 = w_2 = 2$, $w_3 = .7$, $w_4 = .8$, $w_5 = .3$, and $w_6 = .4$. When the VLR is outside the MSC, the weights were chosen to be $w_1 = w_2 = 1.5$, $w_3 = .7$, $w_4 = .7$, $w_5 = .15$, and $w_6 = .2$. (These represent plausible values. Real values, in current systems, depend on the specific switch architecture.)

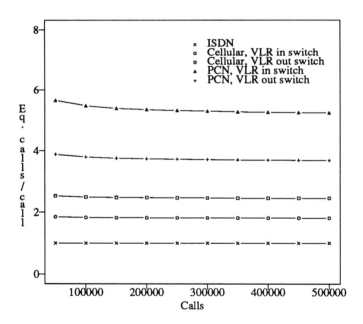

Figure 3. Switching Costs for ISDN, Cellular and PCN

For these example parameters, the signaling network impact of the different network configurations is clearly more important than the switching impact. Of course, the results are extremely sensitive to the chosen parameters. This highlights the importance of carefully estimating these parameters for specific applications.

5. Conclusions

We have analyzed the signaling load associated with terminal mobility in PCN. The number of SS7 bytes per wireless call has been derived for a PCN network using the GSM protocol. For an extremely dense outdoor center city environment with pedestrian traffic, we compared PCN, cellular and ISDN. For this example, we observed that the additional burden placed on the SS7 network is 4 - 11 times greater for cellular than for ISDN and 3 - 4 times greater for PCN than for cellular, depending on whether the VLR is located inside or outside of the MSC.

Methodology for determining the relative switching cost of a wireless call compared to a non-wireless call has also been derived. For the example of an extremely dense center-city environment, the relative increase in the required switch capacity is less than the relative increase in signaling load. Although our quantitative results are not general, that result is probably qualitatively valid.

All of our results are for illustrative purposes only. The results are very sensitive to assumptions on mobility and on the network configuration, in particular the cell size and location area size. This highlights that careful investigation of these parameters is needed for specific applications. Further investigation should also be made into alternative strategies for procedures such as location updates that place a heavy load on the signaling network [10].

6. Acknowledgement

The authors are grateful to Gregory Pollini for his comments and careful reading of the manuscript.

References

[1] Audestad, Jan A. "GSM General Overview of Network Functions" *Proc. International Conference on Digital Land Mobile Radio Communications,* Venice, 1987.

[2] Audestad, Jan A. "Network Aspects of Public Land Mobile Networks Application of Signalling System No. 7," *Proc. 3rd Nordic Seminar,* Paper 5.1. Copenhagen, September 1988.

[3] CCIR Report M/8 (mod. F), "Future Public Land Mobile Telecommunication Systems," Doc. 8/1014-E, December 1989.

[4] ETSI-European Telecommunications Standards Institute GSM Recommendations

[5] May, Adolf D., **Traffic Flow Fundamentals,** Prentice Hall, Englewood Cliffs, NJ, 1990.

[6] Meier-Hellstern, K. S., Alonso, E., O'Neil, D. R., "The Use of SS7 and GSM to Support High Density Personal Communications," WINLAB, Rutgers University Technical Report 24, December 1991.

[7] Meier-Hellstern, K. S., Alonso, E., "Signaling System No.7 Messaging in GSM," WINLAB, Rutgers University Technical Report 25, December 1991.

[8] Meier-Hellstern, K. S., Alonso, E., Pollini, G. P., "Computation of Location Area Update and Handover Rates for Cellular and PCN Systems," WINLAB, Rutgers University Technical report in preparation.

[9] Minhas, H. "GSM Signalling on the Radio Path: A Dimensioning Study," *Proc. International Conference on Digital Land Mobile Radio Communications,* Venice, 1987.

[10] Okasaka, S., Onoe, S., Yasuda, S., Maebara, A., "A New Location Updating Method for Digital Cellular Systems," *Proc. 41st IEEE Vehicular Technology Conference,* May 19-22, 1991, St. Louis, MO, pp. 345-350.

[11] Robrock, Richard B. "The Intelligent Network - Changing the Face of Telecommunications," *Proc. of the IEEE,* vol.79, No.1, January 1991.

[12] Seskar, I., Marić, S. V., Holtzman, J., Wasserman, J., "Rate of Location Area Updates in Cellular Systems," *Proc. VTC '92,* Denver, CO, May 1992.

[13] Thomas, R., Gilbert, H., Mazziotto, G., "Influence of the Movement of the Mobile Station on the Performance of a Radio Mobile Cellular Network," *Proc. 3rd Nordic Seminar,* Paper 9.4. Copenhagen, September 1988.

[14] "The Role Of Signaling System No. 7 in the Global Information Age Network," **IEEE Communications Magazine,** July 1990, Vol. 28, No. 7.

A Microkernel for Mobile Networks

Probal Bhattacharjya WINLAB	Souripriya Das WINLAB	B . Gopinath Electrical and Computer Engg Dept.	D. Kurshan Telecommunications Consultant
Rutgers University	Rutgers University	Rutgers University	
Piscataway,NJ 08855	Piscataway, NJ 08855	Piscataway, NJ 08855	Seabright, NJ 07760

1 Introduction

In mobile networks the components of the application systems can move around in the network. Such movement causes change in system parameters which may violate the assumptions on which the system was working thus far. Such violations are detected and then active components of the system collaborate among themselves to take appropriate corrective actions so that service semantics can still be maintained.

In general, services on mobile networks are mobile distributed applications. Distributed applications can be naturally modeled in a concurrent object model. Objects correspond to the distributed active components or entities, and the interaction among objects corresponds to coordinated communication and collaboration among the distributed active components. An operating system that provides support for a powerful concurrent object model can be used as a platform for developing distributed applications. We developed a model for concurrent objects, CO* [1], that supports concurrent and composable objects, and coordinated communication among groups of objects through a process of agreement. We have designed and prototyped a microkernel for an operating system, WIN*OS that directly supports the CO* model. Development and maintenance of services for mobile networks will be facilitated by the availability of the microkernel and other mobility-specific services implemented on top of the microkernel.

Blackboard, an example of such application, demonstrates the power of the microkernel. It runs simultaneously on several personal computers interconnected by a wireless network. Each computer's screen contains a private area for note taking and a public area for blackboard communication with other computers. A user can join in or leave a session at any time, or move his work from one computer to another. The application, developed on top of the microkernel, takes care of the context and the environment of the user.

In wireless communication systems, the mobile units and the base stations are the distributed active components. The mobile units communicate with the base stations through wireless medium. The base stations are interconnected by wired network. When a mobile unit moves from one cell to another, the base station for the new cell has to take care of the context of the mobile unit. This can be done in a way in which the blackboard application takes care of the context of the user when he moves his work from one computer to another.

The rest of the paper is organized as follows. In section two we compare distributed operating systems with single machine operating systems. In section three we discuss advantages and disadvantages of the microkernel approach compared to the monolithic kernel approach. In section four we summarize the features of some of the most relevant distributed operating systems that are currently available. In section five we describe the CO* model for concurrent objects. In section six we briefly outline some features of the microkernel for the WIN*OS operating system. In section seven we describe a few applications that we have developed on top of the microkernel. In section eight we conclude with a summary of the paper and some indications about our future work.

2 Distributed Operating Systems

Although there has been a consensus as to how essential distributed operating systems are, there is no precise definition of what a distributed operating system is. In this section we discuss what a distributed operating system is, its purpose, and compare it with single machine operating systems.

Single-machine operating systems were developed to do mainly three things. Firstly, they were used to provide a model of the hardware resources that was easier and more convenient to use. Secondly, resources had to be managed so that they could be used in an efficient and fair manner. Finally, resources had to be protected from possible unauthorized and/or illegal use. The principal resources in such systems were a single CPU, memory, and I/O and other devices.

With the advent of machines with multiple CPUs, single shared memory or shared multiple memory blocks, and I/O and other devices the operating system had to provide more functionality. In case of multiple memory blocks the CPUs and the memory blocks are connected by a fast interconnect. Any memory block is addressable from any CPU. This architecture is commonly known as a shared-memory multiprocessor architecture. The operating system code for such a machine could be running on multiple CPUs simultaneously and hence appropriate synchronization has to be built-in in the operating system code for consistent access to devices and memory.

Finally, modern systems consist of group of distinct machines, each of which could be a single-CPU machine or a shared-memory multiprocessor, connected by a communication network. In such an architecture the operating system has to manage an additional resource namely, the communication network. In such a setup the resources to be managed by the operating system are distributed across the network. The operating system has to provide a model for network-wide resources. As in the case of single multiprocessor machine operating systems, the operating system code would be running on different machines and would, therefore, have to synchronize before accessing devices. But the additional difficulty involved in this case is that continuous dialog may have to be maintained between two or multiple operating system codes during access to devices. This kind of operating systems can, in general, be called distributed operating systems.

The issue of transparency of distribution creates a broad spectrum of distributed operating systems. Transparency of distribution is measured as the extent to which users of a distributed system are or need to be aware of the distributed nature of the system. At one extreme the operating system code may simply provide facilities to access remote machines and nothing else, and on the other extreme it could present the group of machines as a virtual uniprocessor machine to the user. The extent of transparency is often used as a qualitative measure of "distributedness" of a distributed operating system.

Another issue that comes up in the nature of distributed operating systems is that of heterogeneity. The operating system codes that run on different machines can be quite different, as long as, they know how to talk with one another, and the concepts implemented by the different codes are interconvertible.

3 Microkernel Approach

A recent trend in operating systems research and development has been to develop the operating systems as an extendible set of user-level servers built on top of a minimal microkernel, instead of a big monolithic kernel that includes all the servers within itself.

The microkernel approach has several advantages. Firstly, since the servers are implemented at the user-level, it is easy to put in additional servers that can provide new services or an alternate implementation of an existing service. Secondly, a smaller kernel increases the reliability of the kernel code. A disadvantage of this approach, however, is that it may lead to execution time overhead because the server codes cannot assume any of the special privileges that the kernel code enjoys. The performance loss (hopefully minor) is probably well-compensated by the gain in reliability and flexibility. Also with sufficient experience in writing distributed applications, we may later be able to identify, in a domain-specific way, some critical services that may need to be moved into the kernel. Another possibility is to be able to dynamically put a server into the kernel and take it out. That solution, however, may need careful design of the server code.

Our approach has been to design a microkernel that provides only the essential services and nothing else. After reviewing a number of current distributed operating systems and after critically analyzing the basic nature of distributed applications we have identified a set of primitives that is small· and powerful. All other services can be implemented on top of the microkernel.

4 Current Systems

In this section we summarize the features of some of the recent distributed operating systems that are most relevant to our discussion.

Mach

Mach was designed and implemented at Carnegie-Mellon University with support from DARPA during the second half of the 1980s[2]. In 1989 Open Software Foundation announced that it would use Mach as the basis for its new operating system. This had put Mach into the forefront of industry action. A big research and development effort on Mach is still going on at CMU as well as in the industry.

The primitive abstractions in Mach are as follows: task, thread, port, port set, message, memory object.

A **task** is an execution environment for threads. It is the basic unit of resource allocation. A task has protected access to system resources such as processors, port capabilities, and virtual memory.

A **thread** is the basic unit of execution. It is roughly equivalent to an independent program counter operating within a task. All threads in a task share access to all the resources of that task.

A **port** is a one-way communication channel that is logically a queue for messages managed and protected by the kernel. *Send* and *Receive* are the fundamental primitive operations on ports and can be carried out by threads that have the appropriate *port capabilities*. Port capabilities may be sent as messages. At any time only one task can have receive capabilities for a port. Thus the communication primitive is essentially many-to-one.

A **port set** is a group of ports sharing a common message queue. Using port set a thread can receive and service multiple ports. The specific port-id is used to distinguish messages for different ports.

A **message** is a typed collection of data objects that can be sent to and received from ports. Messages are not fixed-size and may contain pointers and port capabilities.

A **memory object** is an item on secondary storage mapped into a task's address space.

The implementation of communication primitives in Mach uses three mechanisms that enhances efficiency: *virtual memory remapping, copy-on-write*, and *copy-on-reference*. To transfer a message between two tasks, the receiving task's memory map is changed to include the memory containing the message in the sending task's address space. If either the sender or the receiver now attempts to write to that shared memory, a new copy has to be made. If messages are between two tasks on separate hosts, then the actual data transfer over the network takes place only when the data is referred to.

The use of ports for reference and the ability to transfer the receive capability for ports introduces a level of indirection between the sender and the receiver. This indirection enables changing of receiver transparent to senders and can be used as a basis for dynamic reconfiguration of communication patterns in an application. A major drawback of the communication primitives is that many-to-many communication is not supported. Implementation of many-to-many using many-to-one primitives is burdensome for application programmers and can make the application inefficient.

Mach only provides synchronization primitives for thread-level synchronization. The primitives *thread_suspend* and *thread-resume* can be used to stop and start threads at appropriate times. No primitives are available for synchronization among threads of different tasks. The *msg-receive* primitive for receiving from ports cannot be used for inter-task synchronization because only one task can have receive capabilities for a port at a time. Such synchronization must be achieved by writing special daemons.

Chorus

The Chorus system was designed and implemented at INRIA, France during 1980–1986 and is currently being developed at Chorus Systemes [3][4]. Chorus provides a small nucleus on top of which a number of subsystems have been designed to support applications that were originally developed for running different existing operating systems.

The basic abstractions implemented by the Chorus nucleus are: unique identifier, actor, thread, message, port, port group, region.

A **unique identifier** is used to name a Chorus objects such as actors, ports, etc.

An **actor** is the unit of resource allocation. It defines a protected address space supporting the execution of threads that share the resources of the actor. Chorus defines three types of actors that have different extents of *trust* and *privilege*: *user actors*, *system actors*, and *supervisor actors*. Any given actor is tied to a single site and cannot migrate to another site.

A **thread** is the basic unit of execution. A thread executes within the execution environment of an actor and shares all the resources of the actor with other threads in the actor. Threads in an actor can run concurrently possibly on different processors in case of multiprocessor machines. Threads can communicate using messages and ports, but threads within the same actor can also communicate using shared memory.

A **message** is a contiguous byte string which is logically copied from the sender's address space to the receiver's address space. Large messages may be transferred efficiently using a *copy-on-write* technique or by simply moving page descriptors. Two message passing schemes are allowed: asynchronous unreliable messages, and reliable remote procedure calls.

A **port** is used for addressing in communication. This decoupling of a service interface and its implementation provides the basis for dynamic reconfiguration. A port is a logical queue of messages. At a time only actor can receive from a port. Ports may be transferred from one actor to another that results in transfer of receiving capability on that port to the latter actor.

A **port group** extends point-to-point message passing semantics by allowing messages to be directed to an entire group of threads. A port can be members of multiple groups. When a port group is created it is empty; ports can be subsequently added and deleted.

A **region** of an actor's address space contains a portion of a segment of virtual memory mapped to a given virtual address with a set of access rights.

It may be noted that the concept of port group in Chorus is different from that in Mach. Port groups in Mach can't be used to send messages to threads in multiple tasks. Port groups in Chorus can be used for many-to-many communication, but lack of any ordering guarantee makes them difficult to use for distributed applications.

Amoeba

The Amoeba distributed operating system has been designed and implemented at Vrije University, Netherlands during the 1980s [5][6][7]. Amoeba is currently being used for systems distributed across several European nations.

The primitive abstractions in Amoeba are as follows: process, thread, process group, capability, passive object.

A **process** is an execution environment for threads and is the basic unit of resource allocation. A process has protected access to system resources such as processors, capabilities, and virtual memory.

A **thread** is the basic unit of execution. It is roughly equivalent to a different program counter operating within a process. All threads in a process share all the resources of that process.

A **process group** is a set of one or more processes. Process group names are used for *ordered group broadcast* communication among all the processes in the process group. The protocol for group communication allows for process/site crash and message loss.

A **capability** is used for the naming and protection of a passive object. It also contains a *port* identifier of the server process that manages the passive object. Capabilities are implemented at the user-level and are protected cryptographically.

A **passive object** is the unit of encapsulation. The state of a passive object can only be accessed through the use of the operations defined in the object. A client process that needs to do an operation on an object sends a request to the server process that manages the object and blocks. The message contains the capability for the object, the operation specification, and any parameters for the operation. After performing the operation the server process sends back a reply that unblocks the client. This client-server communication scheme is called *remote operation*.

The support for ordered group broadcast makes the Amoeba communication primitives attractive for implementing distributed applications that requires group communication. The ordering of the group broadcast and the blocking nature of the communication primitives makes them useful as synchronization primitives also. Although these primitives can implement consistent interleaving or ordering of related actions in programs distributed across a network, simultaneity of such actions is difficult to achieve.

ISIS

The ISIS toolkit is a distributed programming environment developed at Cornell University during mid and late 1980s [8][9]. ISIS has been developed on top of Unix and it is currently in use at several hundred locations worldwide.

Two aspects of ISIS are key to its overall approach: virtually synchronous process groups, and reliable multicast protocols.

A **virtually synchronous process group** consists of a set of processes cooperating to execute a distributed algorithm, manage replicated data, provide a service in a fault-tolerant manner, or otherwise exploit distribution. Groups can be overlapping and group membership can change dynamically. ISIS supports four types of groups: a *peer group* is a group of "equal" processes, a *client/server* group is a peer group of processes act as servers for a potentially large set of client processes, a *diffusion group* is a type of client/server group in which servers broadcast messages to the full set of servers and clients, and a *hierarchical group* is a tree-structured set of groups in which a *root* group maps the initial connection request to the appropriate subgroup and the application subsequently interacts with this subgroup.

A collection of **reliable multicast protocols** are provided for use by processes for group interactions. Reliability in ISIS encompasses *failure atomicity*, *delivery ordering guarantees*, and a form of *group addressing atomicity* that makes membership changes synchronous with group communication. A *virtual synchrony* property is supported by the multicast protocols, that is, even when operations are performed concurrently and multicasts are issued asynchronously, algorithms can still be developed and reasoned about using a simple, synchronous model.

The use of dynamically changeable process groups and reliable multicast protocols with guaranteed delivery ordering makes ISIS an attractive programming environment for distributed applications development. As in the case of Amoeba although these communication primitives can be used to implement consistent interleaving of related actions across distributed programs, the notion of a distributed step with logically simultaneous actions is difficult to achieve.

Clouds

Clouds is a distributed operating system developed at Georgia Institute of Technology during the second half of the 1980s [10][11]. It runs on general purpose computers connected by a local-area network. The system is composed of *compute servers* for execution of threads, *data servers* for storage and retrieval of persistent data, and *user workstations* for user-interaction. The structure of Clouds is a three-level hierarchy: minimal kernel that supports mechanisms for memory and processor control, trusted system objects for low-level operating system services, and application objects that provides high-level services.

There are two primitive abstractions in Clouds: passive object, and thread.

A **passive object** is a persistent virtual address space. The state of an object is accessible only through the use of the code in the object. Any entry-point of an object can be invoked from outside. Arguments are passed into the object with invocation and results are returned to the invoker of the entry-point when the invocation returns. The arguments and results must be data, and not addresses. Each object has a global system-level name called a *sysname*. A sysname is a globally unique bit-string. User's can assign high-level names to objects that are translated to the sysname at run-time. Due to the overhead associated with invocation and storage, objects must be somewhat "heavyweight".

A **thread** is the basic unit of execution and resource allocation. Threads are created by an interactive user or under program control. A thread is not bound to a single address space. When a thread invokes an entry-point of an object the thread can access the persistent data in the object, and may invoke operations on other objects. Multiple threads can simultaneously enter a single object. Concurrency control needs should be handled by programmers of objects through system-supported synchronization primitives such as locks or semaphores.

Shared objects may be used for communication among threads. Implementation of distributed steps, where every participant needs to agree that a step is over, is difficult due to the lack of direct thread to thread communication. Polling of shared objects may be used for that purpose, but that solution has its disadvantages.

5 CO*: A Concurrent Object Model

A distributed application can be naturally modeled in a concurrent object model. Just like objects, processes in a distributed application interact with one another to implement the properties of the application. We have developed a model of concurrent objects, CO* [1], that is supported by the operating system WIN*OS. The CO* model allows both specification and implementation of distributed applications to be specified in the same model. This reduces the semantic gap between specification and implementation which has been a persistent problem in distributed software development. In this section we briefly describe the salient points of the CO* model.

Objects: An object is characterized by a *behavior* and a set of group names. The behavior of an object specifies the way the object reacts to changes in its environment. The behavior of an object could dynamically change in response to changes in the environment. Objects are *internally concurrent* and *composable*. Object behaviors can be specified that can achieve arbitrary *relationships* among objects, e.g., dynamic multiple inheritance can be easily achieved.

Groups: An object can be member of multiple groups. There is a many-to-many relationship between objects and groups. Groups can be created/deleted and group membership can change dynamically. Group names are used to refer to all objects in a group and this mode of referring to the object suffices so that objects are not required to have unique identifiers.

Group-steps: Each group goes through a sequence of steps. Step sequences of different groups are independent of one another unless object behaviors impose some relationship. The evolution of a computation is thus characterized by the step sequences of all groups.

Agreement: A group-step can go through only when every object that is currently a member of the group agrees that the step can go through. Each group step is therefore a *distributed coordinated step*. Depending upon possession of different privileges, different object members of a group can exercise different extents of control on the process of arriving at an agreement. The process of agreement achieves *protected distribution of information in a synchronized manner.*

Group Hierarchies: It is possible to relate multiple groups in a hierarchy dynamically. Such a group hierarchy creates an agreement hierarchy for the different group-steps.

6 The CO* Microkernel for WIN*OS

Support for the CO* model is provided directly from the microkernel. The microkernel virtualizes the network into a distributed, synchronized, active memory. The main abstraction supported by the microkernel is called *harmonic clocks*. In this section we describe the semantics of the harmonic clock abstraction.

A harmonic clock evolves through a sequence of steps. A step of a harmonic clock can be thought of as a distributed AND-line. As in a distributed AND-line every object tuned to the clock must agree that the clock can advance to the next step. Only those objects that are tuned to a clock can control the advance of the clock. Clock names are unique throughout the network and hence location independence can be easily achieved. Clocks also directly support the notion of many-to-many communication and democratically controlled advancement of steps.

The operations that can be performed on clocks by the objects are as follows:

> CREATE: creates a clock.
>
> DESTROY: destroys a clock.
>
> TUNE: tune in to a clock.
>
> DETUNE: tune out of a clock.
>
> INHIBIT: inhibits the advancement of a step of a clock.
>
> RELEASE: frees a clock for advancement
>
> TRIGGER: requests an advancement of a clock.
>
> WRITE: writes information onto the clock.
>
> READ_LAST: reads latest stable information on the clock.

The microkernel decides on the advancement of a step of a clock using the following criterion. If every object tuned to the clock has released the clock for advancement and it has been triggered then the clock should be advanced. All the information written to the clock during the current step are put into a bag and the bag is made available to all objects that are tuned to the clock in the next step. The sequence of bags corresponding to the sequence of advancements of a clock can be used to maintain history.

Apart from the operations mentioned above a number of other operations are defined for the harmonic clock abstraction for the purpose of fault-tolerance. These include: retract, abort, fault.

Not everybody tuned to a clock can do all of the above operations on the clock. A capability-based protection scheme is used to specify and enforce authorization in a distributed manner.

The microkernel supports defining of hierarchy of clocks. Such hierarchies can be used to achieve more sophisticated synchronization schemes not achievable by simple unrelated clocks.

Heterogeneity has been a major issue in the development of the microkernel. It can go under any standard host operating system as a network device driver as long as support for interruptible processes is available from the host operating system.

A number of prototypes of the microkernel — three for personal computers and two on Sparcstations — have been developed to date. A number of distributed and mobile applications have been developed on these prototypes. The size of the microkernel prototypes have been remarkably small.

7 Applications

A number of distributed applications have been developed on top of prototypes of the microkernel: distributed games that allow mobility, simulation and dynamic configuration of networks from executable specifications, distributed resource allocation, etc. In this section we briefly outline two applications: distributed mobile games, distributed resource allocation.

Mobility Management

The phenomenon of mobility is ubiquitous in distributed computing and communications applications in general, and specifically in wireless and cellular communications applications. Management of mobility is, therefore, critical in distributed applications.

In a mobile application a number of system components may need to move. Client programs may need to move because clients move. Server programs may need to move because changes in load pattern may require them to move close to the high-load area. Devices such as telephones, printers, fax machines etc. may need to move due to physical constraints arising in the system. Databases may need to move for security or load balancing reasons.

The semantics of all services must be maintained despite mobility, that is, the clients or users of the services should be completely unaware of mobility in the system. Loss or resequencing of data must be prevented. Another problem caused by mobility is frequent changes in load pattern. The system should be able to respond to the changes in load pattern by making appropriate changes.

A system works on top of a set of axioms that are satisfied by the system parameters. Movement of components may change the system parameters in a way that the axioms are not satisfied any more. At this point the system parameters have to be readjusted so that axioms are satisfied once more before normal operations can resume. Thus it comes down to the of detection of the changes that cause violation of axioms and then taking appropriate corrective actions by components that are distributed across the system.

Developing such applications requires strong support for synchronization, communication, and protection among distributed components. The microkernel for WIN*OS provides such support and WIN*OS provides services that are specifically geared towards mobile applications. This makes mobile applications easier to develop and manage. We have developed a few prototypical games over wireless networks that allow mobility of players.

Distributed Resource Allocation

The resources (such as different devices, links) in a distributed system are scattered all over the network. Getting a service from such a system usually involves getting access, possibly exclusive, to a number of such resources at the same time. Allocation of resources to service requests must be done in a consistent manner so that the sanctity of the individual resources is not violated.

There are two different approaches for solving the resource allocation problem: centralized and distributed. In a centralized approach there is one resource manager that makes all the decisions about allocation of resources. This solution has several disadvantages. The single manager is a bottleneck in the system. The system is not scalable because with increased the number of resources the performance goes down considerably. Also the system is unreliable because the resource manager may crash bringing the whole system down. The distributed approach on the other hand is both scalable and reliable. In this case different resources are managed by different resource managers. The resource managers collaborate among themselves to decide on allocation of resources they manage.

A distributed resource allocation problem may be specified as follows. There is a set of resources. Each resource is managed by one or more resource managers. Services are defined to require subsets of the set of resources. Clients request for services. The goal is to allocate resources to services so that sanctity of resources is not violated, and progress is guaranteed if resources are available. All resource managers are identical except for possibly some identifications.

The solution for this problem requires collaboration among the resource managers. The support for coordination in the microkernel facilitates both the specification and implementation of such collaborative activities. We have implemented a version of this solution on top of the microkernel.

8 Conclusion

We reviewed a number of currently available distributed operating systems and pointed out some of their shortcomings. We described the CO* model of concurrent objects. Next we outlined the main features of the microkernel for WIN*OS. Finally we described two of the applications that we have developed on top of the microkernel: distributed games that allow user mobility, and the distributed resource allocation problem.

A number of prototypes of the microkernel have been developed so far both for personal computers and for sparcstations. A number of applications have been developed on these prototypes. We plan to develop new services that are specific to mobile applications on top of the microkernel.

9 Acknowledgments

We would like to thank Sanjoy Mukherjee for implementing the prototypes of the microkernel and for the long discussions we had about the semantics some of the microkernel primitives. We thank Weining Wang for reviewing a number of current distributed operating systems.

Bibliography

[1] Souripriya Das, B. Gopinath, D. Kurshan. *Foundations of Concurrency among Objects.* In IEEE COMPCON, pages 543–549, February, 1991.

[2] M. Acetta et al. *Mach: A New Kernel Foundation for UNIX Development.* In Proceedings of the Summer 1986 USENIX Conference, pages 93–112, July, 1986.

[3] M. Rozier et al. *Overview of the Chorus Distributed Operating Systems.* CS/TR-90–25, Chorus Systemes, 1990.

[4] M. Rozier et al. *Chorus Distributed Operating Systems.* Computing Systems, pages 305–370, Fall, 1988.

[5] Andrew S. Tanenbaum et al. *Experience with the Amoeba Distributed Operating System.* CACM, pages 46–63, December, 1990.

[6] M. Frans Kaashoek, Andrew S. Tanenbaum. *Group Communication in the Amoeba Distributed Operating System.* In Proceedings of the 11th International Conference on Distributed Computing Systems, pages 222–230. IEEE, 1991.

[7] S. J. Mullender et al. *Amoeba — A Distributed Operating System for the 1990s.* IEEE Computer, May, 1990.

[8] K. Birman, A. Schiper, P. Stephenson. *Lightweight Causal and Atomic Group Multicast.* Tech Report, Cornell University, May, 1991.

[9] K. Birman. *The Process Group Approach to Reliable Distributed Computing.* Tech Report, Cornell University, July, 1991.

[10] P. Dasgupta, Richard J. LeBlanc Jr. *The Clouds Distributed Operating System.* IEEE Computer, 1991.

[11] P. Dasgupta et al. *The Design and Implementation of the Clouds Distributed Operating System.* Computing Systems, pages 7–46, Winter, 1990.

Querying locations in wireless environments[1]

T. Imielinski and **B. R. Badrinath**
Department of Computer Science
Rutgers University
New Brunswick, NJ 08903
e-mail:{imielins,badri}@cs.rutgers.edu

Abstract

Tracking mobile users and answering queries about their locations are new and challenging database problems. Preliminary
solutions to two of the major problems, control of update volume
and integration of query processing and data acquisition, are discussed along with the simulation results. Further, new challenges
to the database field brought about by the rapidly growing area of
wireless personal communication are presented.

1 Introduction

The rapidly expanding technology of cellular communications will give
users the capability of accessing information regardless of the location
of the user or of the information. It is expected that in the near future,
tens of millions of people in the U.S. alone will carry a lightweight, inexpensive terminal that will give them access to a worldwide information
network called PCN (Personal Communication Network). These users
will be constantly relocating between cells of size much smaller than today (future cell size might be a building or a floor of a building). Thus,
within a day, a mobile user may cross the boundaries of these small cells
tens and possibly hundreds of times.

Mobile users will be able to ask ad hoc queries addressed to large
databases describing the local area: information about people, places,
its geography, services available, etc. Typical queries from mobile users
will range from simple requests such as "Where is X?" or "Where is the
nearest doctor?" to more complex "Find me the the the best route to the
hospital with the best traffic conditions." Queries may also request data
from a mobile source in a location *transparent* mode: the recent sales
figures from a traveling salesman who stores this data in the memory of
his mobile palm-top terminal.

Before this vision can be fully realized, a number of new database
research problems have to be solved. These problems include the control
of massive location updates, the integration of querying and communication, and the partition of knowledge across the mobile network. It is
estimated (Meier & Hellstern et al., 1991) that the additional signalling
load generated by PCN will be 4-11 times higher for cellular networks
than for ISDN and 3-4 times higher for PCN than for cellular networks
themselves. Most of this traffic will be due to location updates from
constantly relocating users. Additional traffic, which is difficult to predict yet, will be due to the massive data transfers generated by ad-hoc
queries.

The goal of this paper is to present a number of new data management problems hopefully of interest to the database community that
arise due to the *mobility* of both data sources and data consumers. The
management of mobile data is an important part of the emerging area
of *mobile or nomadic computing*. It can also be viewed as an integral

---

[1]This is an expanded version of the paper in VLDB[4]

component of a broader vision of universal and worldwide access to information independent of its form, representation, *location* and *mobility* of users. In addition to the discussion of the new problems, we will also provide some preliminary solutions and simulation results. The full body of results, due to lack of space, will be presented elsewhere.

The work presented in this paper is part of a larger project called DataMan which addresses all of the above questions. The DataMan (a logical successor to WalkMan and WatchMan) will be a distributed knowledgebase capable of handling a large number of queries from hand held terminals[2]. Such a system running on small pocket terminals can replace current pocket organizers by making it possible to access and to process information at *any* location and at *any* time. Information could include maps and navigation, local services, and finally identification of other moving objects as well. The DataMan project will be described elsewhere in more detail.

This paper has two goals:

- Provide preliminary solutions to two of the major problems mentioned above: control of the update volume and the integration of query processing and data acquisition.

- Demonstrate new challenges to the database field brought about by the rapidly growing area of wireless personal communication. Here we present a number of open problems and challenging research questions addressed to the database community.

In Section 2, we first present a general architecture of the cellular system. The next two sections cover in detail the issues of update and query processing. These sections contain the main technical contributions of the paper.

In Section 5, we outline a number of important open questions and future direction of this work. Section 6 summarizes related work and finally, conclusions are presented in Section 7. Preliminary simulation results are described in the appendix.

2 General Architecture

The system consists of a fixed information network extended with a wireless component. The wireless elements include: wireless terminals, base stations and switches. The whole geographic area is partitioned into *cells*. Each cell is covered by a *base station* which is connected to the fixed network and provides a wireless communication link between the mobile user and the rest of the network. The allocated bandwidth of radio spectrum[13] in North America includes frequencies between 870 - 890 kHz (from base station to mobile) and 825-845 kHz (from mobile to base). This bandwidth is subdivided into 666 channels which are shared and reused among base stations which are sufficiently far apart. Currently the average size of a cell is of the order of 1-2 miles in diameter[19]. It is expected, however that with a new generation of wireless systems this size will shrink to possibly 50-200 meters to accommodate a much larger set of users and still benefit from bandwidth sharing and reuse schemes.

---

[2]The prototype of DataMan will make use of current resources of the Wireless Information Network Laboratory (WINLAB) at Rutgers University which is supported by 18 industrial research sponsors. These resources include a complete cellular communication system with base stations and switching software. We will assume the availability of a small portable low-powered terminal with limited memory (say 8-16MB) which can be switched on and off by software (to minimize power consumption).

Additionally, we will assume that there is a hierarchy L of location servers which are connected among themselves and to the base stations by a standard network (Figure 1). Typically location servers correspond to Mobile Switching Offices and there are about 60-100 base stations "under" a location server. Location servers are responsible for keeping track of addresses of users who are currently residing in the area "below" the location server. These addresses may be stored as exact locations - i.e., the identifier of a cell the user is currently in, or as approximate locations; a zone, or a partition of cells.

Each user (sometimes called mobile terminal) will be permanently registered under one of the location servers; additionally the user may also register as a visitor under some other location server. Base stations will always be aware of mobile terminals which are active within their cells. Location servers, however, do not have to know in which cell a given mobile terminal is currently located - they can always find out by means of paging – a multicast message sent to a subset of base stations "under" the given location server. Since the location server needs to contact the base stations over the fixed network, the cost of broadcasting is equal to the number of base stations to which paging requests are sent[3].

By the users' location we will understand an identifier of a cell in which the user is currently residing. The database storing users' locations will typically be distributed among many location servers. Importantly, the database will almost never store the actual exact location of a user (i.e., the cell id) but either some outdated previous position or a set of "possible locations." Therefore, it will almost always store incompletely specified location data to avoid massive and frequent location updates.

Querying locations will in general lead to combinations of table look-ups and selective broadcasts. For instance, a simple query "Where is John?" asking for a cell number where John currently resides, depending on the addressing protocol may involve accessing John's home location server and then paging him within base stations (corresponding to cells) belonging to the area covered by this location server. If he is not found here then we will have to find the location server under which John currently resides. Several protocols can be used: e.g., broadcasting from the root of the hierarchy of location servers, following the chain of forwarding addresses (if the user leaves them) etc.

Static and Mobile Databases

Each location server will keep a "home database" of all user related data for users with the permanent residence within the location area administered by the server. Additionally, there will be a visitor database[4] which stores information about users who are not living in the area but are currently visiting the area. Finally, there will be databases related to specialized services such as 1-800-Doctor which would keep information about doctors who are currently residing in some predefined area possibly different from the area covered by the location server. Such specialized databases (views) will be dynamically created as it is the case now for standard telephone services. Databases can also be mobile, especially with the advent of new generations of cellular systems. People

---

[3]The cost of broadcast over the wireless medium does not depend on the number of recipients. However, broadcasting from the location server to the base stations is performed over the fixed network (tree of location servers and base stations). This involves point to point communication from the location server to each of the base stations. Hence, the cost of broadcasting in this case is proportional to the number of base stations.

[4]These terms have been introduced earlier in the context of cellular networks[16]

88

will be carrying pocket terminals and running computer programs such as editors, data acquisition packages, etc. while relocating. In such a situation, location servers, specialized servers, etc., will become mobile sources of information, which others in the network may constantly want to access. For instance, the user may be collecting data while moving, and he may not necessarily report all of this new information back to the static server.

In the case when all components of the network become mobile, users and databases have interchangeable roles: each of them can query the other.

2.1 Queries and Updates

The set of users together with their characteristics (including location) can be viewed as a database distributed between different location servers. We will assume a simple object oriented view: our database will have a form of the class of users characterized by a number of methods and instance variables (attributes). One of these methods will determine the location of the user. This class will be stored as a relation which is horizontally partitioned between different location servers according to the "home areas" of users. Further in the paper we will talk extensively about querying and updating such a database. By updates we will really mean *location updates* of the current location of the user. These massive updates will not be transactional in nature and will not require locking. Queries will be standard relational queries which could involve the location attribute both in selection as well as in the target clause.

In the next two sections we address two fundamental questions: how to update and how to query locations in a mobile environment.

3 "Do not tell me - if I do not want to know": Updates

The main idea in this section is that users do not have to inform the location server about each and every change of their location. In other words we should leave a certain degree of ignorance about the state of the system (i.e. user's locations in this case). We postulate that ignorance should be *bounded*, i.e., we should not be too "far off" from the truth. In fact, the *bounded ignorance*[5] approach will guarantee that the real position of the object and the position which is known are always in the same *partition*. Partitions are defined depending on the user mobility patterns and will help to maintain a certain degree of *knowledge* about users' whereabouts (i.e., "I do not know where you are exactly at this time but I know that you are in New Brunswick"). Maintaining such knowledge comes with a cost, but as we will demonstrate it will be much cheaper than keeping the complete information about individual locations.

3.1 Model of agent's mobility within a location server

Intuitively, partitions decrease the overall cost by "gluing together" cells between which the user relocates very often and by separating cells between which the user relocates infrequently. For example, assume that we have gathered the data about the user's mobility between base stations and determined the frequencies with which the user relocates.

---

[5]This term is borrowed from [15], where it is used in a different framework — allowing a replicated system to violate integrity constraints by a bounded amount

In Figure 2, the numbers on the edges indicate the number of times the user relocates between the end locations connected by the edge during a certain interval of time. Edges with a high crossing rate are contained in the same partition and edges with a low crossing rate in separate partitions. In this way the update cost slightly increases (compared to one partition, when the update cost was simply zero) while the query cost significantly decreases, since we only broadcast to base stations within a partition. Obviously there is a tradeoff. Therefore, there is a need for an optimal partitioning which minimizes the expected cost of querying and updating. This task requires knowing more information about users' mobility and patterns of calls.

Users usually relocate, especially within their home areas according to stable daily patterns (routines). Most of us commute from home to work (and back) along the same route and then have daily routines which are also periodic. The information about each user's mobility is captured by the notion of a *user profile* analogous to credit card and telephone companies storing financial and calling profiles on a per user basis[6]. The user's profile will contain:

1. Probabilities of relocation between any two locations within a location serverbase(1).

2. Average number of calls per unit time

3. Average number of moves per unit time

We assume that the profile is obtained either by monitoring user activity (mobility and frequency of calls) over an extended period of time or/and by getting some of this information directly from the user. In general, of course, probabilities will be approximated by relative frequencies. In the future, we will consider models which allow different patterns of mobility at different time intervals (which is much more realistic, since our daily work routines are different from after hour routines). This model helps us to decide when to notify the location server about the change in our position - it will occur only when we cross the borders of predefined partitions. We assume that partitions will be modified periodically. Shifts in trends of user mobility will be detected and partitions will be modified accordingly at reorganization points. In a way it is analogous to periodic index reorganization.

We will describe several possible algorithms to locate users, assuming some partition based scheme. In the appendix we present preliminary simulation results.

3.2 Algorithms for locating users

In the following discussion we assume that users reside under their home location servers. Thus, the partitions consist of base stations belonging to the same location server. A more general case of user movement across location servers is presented in [3]. We will consider three different strategies within the partition scheme. In each of these strategies, a user informs the location server when he crosses partitions thus incurring update cost, and the search for a user is always within a partition. However, these strategies differ in the way a user is located within the partition. The three strategies are:

1. *Broadcast:* a broadcast is initiated within the partition to locate a user. The cost of locating a user is bounded by the size of the partition, i.e., the number of base stations within the partition.

---

[6]Users' profiles will be distributed according to their home location areas. Additionally, we may also store aggregate profiles for specific geographic regions such as a shopping mall, a train station or a football stadium.

2. *List:* a list of locations within each partition is maintained. This list is sorted by likelihood of being at a given location. Each location is tried successively. Maximum cost will be incurred when a user is in the least likely location within a partition.

3. *Pointer:* here, each time a user moves to another location, within the same partition, a forwarding address or a pointer is maintained at the previous location. When locating, the pointer is followed from the location first visited in the partition to the current location (similar to call forwarding). If the user visits a location which he has visited before we "cut" the resulting loop. The cost of locating a user is bounded by the the length of this pointer chain; at most the number of base stations within the partition. When a user moves to a different partition, a new pointer chain is maintained in that partition. We assume that leaving forwarding addresses (pointers) at base stations has the same cost as informing a base station of a move. However, we have not included the cost of collapsing pointer loops nor the cost of removing pointers in the previous partition.

In all these strategies, the total cost (update cost + locating cost) depends upon a) the partitions and their sizes, b) user profiles, i.e., the base stations visited, and c) the strategy used.

Simulation results comparing these strategies with respect to the total communication cost (not response time) will be presented in the Appendix.

4 "I do not know but can find out" - Queries

The discussion in the previous section indicates that the database will not usually know the precise location of each user. If the precise location is needed by a query some extra work is needed. This leads to a new paradigm of query processing which involves acquisition of new information in the run time of the query. Before discussing queries and their evaluation we will show how to model the notion of error when dealing with fast changing values.

4.1 Modeling fast changing values - living with errors

Location is an example of a fast changing attribute. Other examples include temperature, speed, pressure etc. in some sensory environment. It is costly to maintain rapidly changing quantities. Therefore we have to be prepared to accept erroneous (outdated) values. In such cases we would like to know, however, what is the margin of error, how to find a better, more precise value and finally, what is the cost of such a precise value finding procedure.

Below we will make a fundamental distinction between the *actual* and *stored* value of a dynamic attribute since most often they will not coincide. The notion of partition will help us to define the margin of error. Various user locating methods (such as paging, pointer and probabilistic addressing schemes) will provide an exact value at additional cost. We will describe all these notions with the help of an example.

Example 1
A user named John will be specified as follows:
Object: John
Partitions: $\{Cell_1, Cell_2\}, \{Cell_3, Cell_4, Cell_5\}$
Correctors: Paging with pointer

Predictors: {John is at home, in $Cell_2$ after 6 P.M.}

Additionally (aside from standard attributes such as age, profession, etc), the following methods will be defined on John:

LOC: John.LOC will return the location of John according to the database (which may be outdated and incorrect). For instance, in our case it could be that $John.LOC = Cell_1$

ERR: John.LOC.ERR will return the partition (a set of locations) which is specified by John's profile and contains John.LOC - the stored location of John. If $John.LOC = Cell_1$ then John.LOC.ERR $= \{Cell_1, Cell_2\}$

loc: John.loc returns the actual location of John. This method may involve not only database access but also additional considerable communication in the form of paging and multicasting (so called *corrector methods*).

Correctors will stand for methods which can be used to obtain the exact value of the location, if it is required (that is x.loc). In general, one could have the same correctors associated with all users (this corresponds to one universal protocol of user location) or different correctors associated with classes of users (due to e.g. mobility patterns). For instance, the pointer method may be good for one profile while probabilistic probing superior for another.

Finally, the last element of John's profile illustrates the notion of the "predictor." Location predictor (viewed as a set of rules or a standard procedural method) will compute a *default* location of the user in case the user himself did not notify the location server about *exceptional* behavior. In John's case we stated that at 6 P.M. he is (usually) at home. The location server, at 6 P.M. will try first the user's home to locate him, unless there is a message from the user stating an exception. In such a case, we may have to resort to one of the paging methods.

It turns out that quite often queries can be answered even in the presence of errors therefore incurring no cost resulting from applying correctors. Some other queries may require some of the erroneous data to be corrected. We will discuss this further in the next section.

4.2 Query processing

As we said in the introduction, our queries can range from simple ones such as "where is X" to more complex ones like "Find me a doctor near the campus" or "Find X, Y and Z such that all of them are on the same highway and Y is between X and Z."

These queries differ in the complexity of *location constraints* - i.e., constraints which involve individual location of users. For example, the query "Find me a doctor near the campus" has one non-location based constraint ("doctor") and another which is a unary constraint on location. The query "Find X, Y and Z such that all of them are on the same highway and Y is between X and Z" involves a more complex constraint - a ternary one (between) plus three unary constraints involving individual locations ("on the highway"). Generally, one can get even more complex queries which should recognize "patterns" of moving objects.

The main new problem arising in query processing in the presence of imprecise knowledge about locations of users can be summarized as follows:

How to minimize the communication cost to "find out" the missing information necessary to answer the query?

Without loss of generality we will assume a domain calculus query language. We will concentrate on describing the evaluation strategy for queries written in a "pseudo SQL" form:

SELECT $x_1.loc, \ldots, x_m.loc$

FROM Users

WHERE $x_1.loc = l_1 \wedge \ldots \wedge x_n.loc = l_n)$

$\wedge C(l_1, \ldots, l_n) \wedge W(x_1, \ldots, x_n)\}$

where $C(l_1, \ldots, l_n)$ is an n-ary constraint imposed on locations $l_1, \ldots, l_n$ and $W(x_1, \ldots, x_n)$ is a constraint on individual objects. Users is a name of the class which stores all instances of users in the system.

The query wants to evaluate a predicate over *actual* locations of users, while the only locations available are $\{x_1.LOC, \ldots, x_m.LOC\}$ with the associated "error" determined by ERR.

A naive strategy to process the query presented above is as follows:

Find objects $a_1, \ldots, a_n$ such that $W(a_1, \ldots, a_n)$ is satisfied. Determine exact locations $a_1.loc, \ldots, a_n.loc$ of each of $a_1, \ldots, a_n$ by paging for each $i(1 <= i <= n)$ the partition $a_i.LOC.ERR$. Check if the result satisfies the constraint $C(l_1, \ldots, l_n)$.

This strategy will, in general, result in too many paging messages. Indeed, paging may be either totally avoided or done only partially. For example, it may turn out that $C(l_1, \ldots, l_n)$ is true *for all* combinations in the cartesian product of $a_1.LOC.ERR, \ldots, a_n.LOC.ERR$. Similarly, the constraint $C(l_1, \ldots, l_n)$ may be false *for all* combinations in the cartesian product of $a_1.LOC.ERR, \ldots, a_n.LOC.ERR$. Finally, we may be able to determine after some partial paging (i.e., to locate a_1 and a_2) that regardless of what the locations of the remaining objects are within corresponding partitions, the constraint is true (or false) saving unnecessary messages. Assuming that the computational costs are much smaller than the communication costs this is a worthwhile gain as the following examples indicate:

Example 2 Give the names of doctors (possibly mobile) located near John's current location.
$\{d: Near(d.loc, John.loc) \wedge Doctor(d)\}$ where $Near(y, x)$ is the predicate which is satisfied when cell x is a neighbor of a cell y (either side by side or diagonal).

Let us assume that the database stores L7 as a location of a doctor d and L12 as John's location. Assume also that u is a partition of the doctor's profile which contains L7 and that v is a partition of John's profile which contains L12. In other words we know that the doctor's real location is somewhere within u (d.LOC.ERR = u) and John's real location is somewhere within v (John.LOC.ERR = v).

There are three possible strategies to answer this query: first page both partitions John.LOC.ERR and d.LOC.ERR and then determine whether the final positions of John and a doctor are "near by." The other two strategies both start by first determining whether paging is required at all: maybe the predicate Near(x, y) is true or false uniformly for all combinations of locations in John.LOC.ERR and d.LOC.ERR respectively. If this is not the case then we have a choice of first paging John.LOC.ERR or paging d.LOC.ERR. These two strategies may lead to different costs.

Notice that paging $u =$ d.LOC.ERR first is clearly advantageous. No matter what the result of paging is, paging of $v =$ John.LOC.ERR is not necessary. Indeed, if John is in l_7 then no matter where v is, the Near predicate is true. If the doctor is in l_3 or in l_4 then no matter where John is the constraint is false. Therefore, the overall expected cost is 3 in this case. This is represented as the first tree in Figure 4, where the nodes labeled by POS and NEG respectively correspond to situations which satisfy or falsify the query constraint. The first of the trees, which corresponds to paging first the location of a doctor, has only one level since no further paging is necessary.

If we page v first then (without selective paging) we still have in each case to page u. This is why the second tree has two levels. Therefore the expected cost is 6 in this case, shown as the second tree in Figure 4 (if we could page selectively it would be 4, since we could just page l_7 and on the basis of u's presence or absence there conclude the truth value of the constraint). If the Error(x, l_{10}) included l_6 (i.e. the partition for v also included l_6) then we could no longer ignore v after paging u.

In general, we are interested in a (expected) minimal communication cost *strategy* to evaluate the query. By a *strategy* we mean the sequence in which we will page partitions (if we page them at all) $x_k.LOC.ERR$. Such a strategy, in general, can be represented by a tree with nodes corresponding to variables $x_i, i = 1, \ldots, m$, and two special *terminal* nodes called POS (positive) and NEG (negative). POS and NEG respectively will indicate that we can already determine whether the constraint is true or false respectively as indicated in our example. The edges are labeled by conditions $(x_i.loc = c_i)$ where c_i belongs $x_i.LOC.ERR$. In the example, we simply labeled the edges by c_i skipping x_i as clear from the context of the picture.

Each node N of the tree is identified with path(N) which is a conjunction of all conditions along the path from the root to the node. A path $(x_{i_k}.loc = c_{i_k})$ for $k = 1, \ldots, m$ terminates with POS node iff the conjunction of all $(x_{i_k}.loc = c_{i_k})$ along the path implies constraint $C(x_1.loc, \ldots, x_n.loc)$. The same path terminates with NEG if the conjunction of all conditions along the path implies that the constraint $C(x_1.loc, \ldots, x_n.loc)$ is false. For instance, in the first tree of our example, all edges labeled by (d.loc = 13), (d.loc = 14) and (d.loc = 17) are terminal nodes. The first two in NEG, the last in POS. Indeed, d.loc=13 implied logically that the constraint Near(d.loc, John.loc) is false no matter what John.loc is (within John.LOC.ERR).

The cost associated with the node is equal to the number of outgoing edges which do not terminate in NEG node (intuitively this is the cost of paging and we do not have to page the locations which we can determine lead to a failure).

The cost of the path is equal to the sum of the cost of nodes along the path. Each path is equally likely (hence we assume that any combination of locations from $x_1.LOC.ERR, \ldots, x_n.LOC.ERR$ is equally likely).

The expected cost of the path is the sum of costs of all paths weighted by the associated probabilities. The expected cost corresponds to the expected number of paging messages which will have to be sent in order to determine if $x_1.loc, \ldots, x_n.loc$ satisfy the query.

How hard is it to find a tree corresponding to the optimal strategy?

In the Appendix we demonstrate that the problem is NP-complete, by showing a strong (and somewhat surprising) connection of this problem and the classification problem in machine learning. A greedy heuristic based on the ID3 [21] like algorithm to construct a tree is demonstrated there.

4.3 Aggregate Queries

Such queries could measure global traffic in the area to help in dynamic allocation of resources such as frequency allocation in a particular cell. For example, the area surrounding a football stadium after a game should be assigned more frequencies since the expected activity may be higher there. Aggregate queries will help in establishing such critical zones. Such queries have to take under consideration the road maps in the areas considered especially when detection of high traffic patterns is of interest. This requires careful *integration* with geographic data, street and building maps etc.

Aggregate queries will involve evaluating aggregate operators such as COUNT, AVG, MIN and MAX over the mobile objects residing in some geographic area. Such queries are very important and common in the PCN environment which we are discussing here. Examples include: "Number of policemen in the stadium area", "Number of Taxi Cabs near the theater", "The ambulance closest to the place of accident" - where "closest" could be defined in terms of some traffic conditions (e.g. number of traffic lights). Finally, more complex queries may inquire about the police car being closet to another moving car etc. Queries of this nature can be formulated both by users (i.e. traffic controllers, taxi dispatchers) as well as by network control management facility (the network may be interested in allocating dynamically more channels to the area with the number of users exceeding a certain threshold, hence there may be lots of aggregate queries formulated in that context).

Such queries will have the following general form:
SELECT AggregateOp(f($x.loc$)) FROM Users
WHERE
$\wedge\ C(x.loc) \wedge W(x)$
where AggregateOp denotes any operator such as COUNT, AVG, MIN, MAX, C is a location constraint and W is a non-location constraint. Examples of such queries are shown in Figures 5,6, and 7.

5 New Challenges

Having shown our approaches to the important problems of update and query in mobile environments, we now proceed to a summary of other challenges and open problems which the mobile environment of personal communication systems brings into the database field.

- Incorporating Data Acquisition into Query Processing

  We have provided here a model for incorporating data acquisition into the query answering process.

  Interesting open problems involve the impact of different types of constraints on the selection process for the best classification tree. It should be much easier to find such a tree if the constraint is a conjunction of unary (binary) predicates. Also, there is a need for a more refined cost model which would also take into account the total *time* necessary to broadcast all paging messages.

  Further, we have ignored the possibility that queried objects may reside in the same area and we may be able to locate a number of them within a single broadcast (we have assumed that finding out about each object requires a separate broadcast).

- New Types of Queries

  A highly mobile and distributed environment creates new scenarios for query processing:

  - Queries depending on location of the person asking the query. Where is the nearest restaurant? Give me the best directions to the hospital including the current traffic reports. Such queries will involve *geographic* and *real time* data. This will lead us to the question of how to create yellow pages dynamically.

— Queries depending on direction of movement Knowing our position on a map will not be sufficient; we may need to know the direction of movement to answer a query about the best way to reach the destination.

- Querying Transient Data

  Information changes so fast in our application that it may change *during* query evaluation. This creates an interesting research problem on the very basic, semantic level. What is the meaning of the answer to a query? How to compute it and how to augment it? How to incorporate prediction into query processing?

  Frequency is another important resource: for example we cannot issue an arbitrary number of paging actions. Usually there is a limited number of paging channels and if all of them are busy we have to wait. This blocking effect has to be treated as part of the general problem of dynamic resource allocation in the cellular network — some of these problems are currently being addressed in [13].

6 Related Work

Cellular technology is rapidly gaining public acceptance. Wireless computing and access to information in a wireless network will soon follow. The research work described in this paper is, to our knowledge, one of the few efforts that is addressing issues of information access in a mobile distributed environment.

Current efforts in providing support for wireless computing have focussed on adapting current cellular technology to hand held computers or wireless terminals. Networking software designed for internet use is not applicable to the wireless network problem of dynamically changing addresses; "seamless" handoffs from one network to another require new solutions. Designing network software that is transparent to movement across networks, as well as the design of hardware and user interfaces for hand held computers, is the focus of research at Columbia [8]. Solutions to maintaining a file system for mobile users (where local disks and excessive amounts of RAM are not feasible), and allowing remote paging from file servers in a mobile environment, are presented in [11, 12].

Automobile navigation systems also provide limited information about surrounding areas by indicating destinations on a display map. This helps drivers to navigate in unknown areas[5]. Continuous monitoring of moving objects for air-defense coverage was the goal of the SAGE project [18]. Continuous monitoring of moving objects (platforms) by means of active objects [7] has been discussed in [6] Here, the cost of polling for the evaluation of conditions is compared to that of monitoring by means of active objects or triggers.

Cellular networking technology for access to future wireless networks (third generation and beyond) is being investigated at WINLAB in Rutgers[13, 14]. Research is focused on network protocols, cellular packet switching, network arrangements, and spectrum reuse. Further, mobility requires that users and service providers need to be located. The issue of distributed location service for tracking mobile users has first been addressed in [1, 2]. Here, the directory server supports find and move operations. A hierarchical directory structure that minimizes the cost of a sequence of find and move operations is presented. Further, on a move, only nearby directories are updated, so that only approximate information about users' location is maintained. For find operation, nearby users will be able to find the exact location but users far away will have to follow a chain of pointers to find the exact location.

Thus, most of the the current research work has concentrated on cellular networking and architectures for hand held terminals in mobile networks. To our knowledge the research work presented here is the first attempt to address the issues of querying information in wireless and mobile environment (in a much wider context than automobile navigation systems).

7   Conclusions

The vision of universal wireless personal communication offers new challenges to the database field. In this paper we have identified a number of such new research problems arising due to mobility. We provided preliminary solutions to update control and query processing problems ; we have also described a number of additional research issues.

This research effort can be viewed as an integral component of a broader vision: universal and worldwide access to information independent of its form, representation, *location* and *mobility* of users. Work on heterogeneous database systems addresses the first two issues; Our focus is on the other two issues — location and mobility. All these features have to be fully addressed in order to achieve the final goal of *total transparency* of access regardless of form, representation, location and mobility.

Acknowledgments

We are grateful to Irv Traiger for the extensive comments and help in improving the presentation of the paper. We would like to thank David Goodman for letting us know the state of the art in cellular technology and Greg Pollini for explaining the models used in cellular networks. Thanks also go to Phil Bohannon, Asher Mahboob, Miles Murdocca, and Wojtek Szpankowski for various comments on the paper.

References

[1] Baruch Awerbuch and David Peleg, "Concurrent online tracking of mobile users", Proc. ACM SIGCOMM Symposium on Communication, Architectures and Protocols, October 1991.

[2] Baruch Awerbuch and David Peleg,"Sparse Partitions", Proceedings 31'st IEEE Symposium on FOCS, October 1990, St. Louis.

[3] T. Imielinski and B. R. Badrinath, "Adaptive tracking of mobile objects," In preparation.

[4] T. Imielinski and B. R. Badrinath, "Querying in Highly distributed environments," Proc. of the 18th VLDB, Auguts 1992, pp. 41–52.

[5] James L. Buxton, Stanley K. Honey, Wadym E. Suchowerskyj, and Alfred Templehof, "The Travelpilot: A Second Generation Automotive Navigation System," IEEE Transactions on Vehicular Technology, Vol. 40, NO. 1, February 1991.

[6] Sharma Chakravarty and Susan Nesson, "Making an Object-Oriented DBMS Active: Design, Implementation, and Evaluation of a Prototype," International Conference on Extending Database

Technology, Venice, Italy, March 1990. (Also in Lecture notes in Computer Science, Vol. 416, pp. 393–406.

[7] Umesh Dayal, et al., "The HiPAC Project:Combining Active databases and timing constraints," Proceedings of SIGMOD Record, March 1988.

[8] Daniel Duchamp, Steven K., Gerald Q. Maguire, "Software technology for wireless mobile computing", IEEE Network Magazine, November 1991.

[9] Garey and Johnson,"A guide to NP complete problems," W. H. Freeman, 1979.

[10] Hyafil, L., and Rivest, R.L., "Constructing optimal binary decision trees is NP-complete", Inform. Process. Lett. 5, No.1, 1976 pp. 15-17.

[11] Bill N. Schilit and Dan Duchamp "Adaptive remote paging for mobile computers" Columbia University, Technical Report, CUCS-004-91, February 1991.

[12] Carl D. Tait and Dan Duchamp, "Service interface and replica management algorithm for mobile file system clients," Proceeding of the Parallel and distributed information systems conference, December 1991.

[13] David J. Goodman, "Trends in Cellular and Cordless Communications" IEEE Communications Magazine June 1991.

[14] David J. Goodman, "Cellular Packet Communications" IEEE Transactions on Communications, VOL. 38. NO 8. August 1990.

[15] N. Krishnakumar and Arthur Bernstein "Bounded Ignorance in distributed systems" In Proc. ACM PODS, Denver, CO, May 1991, pp. 63–74.

[16] Kathleen S. Meier-Hellstern, Eduardo Alonso, and Douglas Oniel, "The Use of SS7 and GSM to support high density personal communications," Third Winlab workshop on third generation wireless information networks, April 1992, pp. 49–57.

[17] Readings in Distributed AI, Edited by Alan Bond and Les Gasser, Morgan Kaufmann, 1988

[18] John F. Jacobs, "SAGE Overview," Annals of the History of Computing, Vol. 5, NO. 4, October 1983.

[19] William Lee "Mobile cellular Telecommunication systems," Mcgraw-Hill, 1989.

[20] C. N. Lo, R. S. Wolff and R. C. Bernhardt, "An estimate of network database transaction volume to support universal personal communication services," Submitted to the 1st International conference on Universal Personal Communications.

[21] J. R. Quinlan "Induction of decision trees", Machine Learning, 1986, pp. 81–106.

[22] John Mylopolous, Alex Borgida, Mathias Jarke, and Manolis Koubarakis, "Telos: Representing knowledge about information systems" ACM TOIS, Vol. 8, No. 4, Oct 90 pp. 325-360.

[23] M. Stonebraker and L. Rowe, "The Design of POSTGRES," Proc. of the ACM SIGMOD conference on Management of Data, Washington, D.C., May 1986.

[24] Gio Wiederhold "Mediators in the architecture for future information systems" Unpublished Manuscript.

[25] Ouri Wolfson and Sushil Jajodia, "Distributed Algorithms for Dynamic replicated data" To appear in ACM PODS, June 1992.

A   Simulation studies

Preliminary simulation studies have been conducted to study the effects of using user profiles, in conjunction with the idea of partitioning, on the cost of tracking users by various algorithms described in Section 3.2. The objective of the simulation study is to evaluate various design choices and not an in-depth performance evaluation.

A.1   Simulation model

In the simulation model we assume that our queries are simple "calls" to locate a user. Updates are determined by mobility. The ratio of queries to updates, termed the call to mobility ratio, is a critical parameter in our studies.

Further, the important aspect of the performance model is the use of profiles in evaluating various algorithms for locating a user. Since the user profiles vary considerably, it is necessary to consider typical profiles and study how each strategy employed to locate a user performs for these profiles.

There are a fixed number of base stations among which users move. Moves and calls are generated according to random variables. The mean times between successive calls or successive moves are given by exponentially distributed random variables. The call/mobility ratio is the ratio of the rate (reciprocal of the mean) of calls to the rate of moves. This ratio is an experimental parameter. When a move is generated, the user either decides to stay in the same base station or moves to the next base station which is determined from the user profile. This profile is represented as a transition probability matrix. This matrix gives the probability of a user moving from base station i to j. From the transition probability matrix, it is possible to determine the number of times a user moves from base station i to j within a specified time interval. Thus, within a time interval, it is possible to determine the weights of the edges, i.e., the number of crossings per time interval. Examples of user profiles, with transition probabilities shown as labels on the edges, are shown in Figure 8. For example, in user profile I (partitions are $P_1 = \{1\}$, $P_2 = \{2, 3, 4, 5\}$, $P_3 = \{6\}$, $P_4 = \{7, 8, 9, 10\}$), the user moves randomly within partitions P_2 and P_4. Profile II (partitions are $P_1 = \{1, 2\}$, $P_2 = \{3, 4\}$, $P_3 = \{5, 6\}$, $P_4 = \{7, 8\}$, $P_5 = \{9, 10\}$), represents the case where a user relocates often between two base stations and then moves to another base station and repeats this behavior.

When a user crosses a partition, the current location is updated, incurring an update cost. The cost of the user informing the base station or location server is more expensive than a base station or location server contacting the user. This is because when a user contacts a base station, there is contention among other users and also the base station needs to authenticate users. Authentication requires messages to be sent between the user and the base station. Hence, more messages and time are needed for a user to contact the base station than the other way around. For purposes of the simulation study, we have chosen the cost of update to be four times that of a call (when the current location is known). This is because of the resource contention due to blocking when communicating from the mobile user to the location server (via the base station). The cost of the call simply reflects the communication cost from the location server to the mobile user. If the location of the user were unknown then the cost would also have to include the cost of search.

Further, a table is maintained in the location server to detect the partition to which a base station belongs. This table gives the partition number for each base station. When a call is generated, the cost of the call, if the current location is not known, will depend upon the strategy employed to locate the user. Thus, for each event, the cost due to either an update or a locate call is accumulated to obtain the total cost incurred by various strategies.

A.2   Simulation results

In this section, we present performance results for strategies under the partition scheme. The performance metric employed in comparing various algorithms is the total cost (update cost + location cost) incurred as a function of the call/mobility ratio. The experiment is repeated for different user profiles. The user profiles chosen are shown in Figure 8.

Figure 9 shows the cost per event (calls + moves) (average of 5 runs, for 100000 events) as a function of the call to mobility ratio for various user profiles. Each curve represents the performance of an algorithm. The various algorithms are: Broadcast, List, and Pointer.

Under high call/mobility ratios, the cost of an event for broadcast converges to the number of base stations in the partition; and to the average length of the pointer chain and the average number of base stations in the partition for the pointer based scheme and the list based scheme, respectively. The list based scheme performs better than both the broadcast and the pointer based schemes for Profile I over a wide range of call/mobility ratios. The pointer based scheme performs better than the list based scheme only for a high call/mobility ratio in Profile II. Pointer based scheme incurs a penalty for informing the base stations of the new address for each move. The lower the mobility, the lower the cost of pointers. Thus under high call/mobility ratios, the pointer based scheme incurs a lower cost than under low call/mobility ratios.

The improvement in performance (decrease in cost) as a result of partitioning is very significant compared to that of broadcasting without partitioning. For example, with the above user profiles and the chosen partitioning, the improvement in performance for moderate to high call/mobility ratios with the pointer and list based schemes is two to three fold over schemes that just employ broadcasting. Further, broadcasting without partitioning (i.e., broadcasting to all base stations under a location server) would perform even worse. The pointer based scheme is suited to users exhibiting random movements within each partition and when the call/mobility ratio is moderate to high (Profile I). Under these circumstances, if a list based scheme is used, then a call will be expensive as one has to try all equally likely locations. However, the list based scheme is suited to profiles exhibiting a more deterministic movement within small partitions and when the call/mobility ratio is low to moderate (Profile II). At low call/mobility ratios (high mobility), the list based scheme does not incur any cost due to updates and when a user is often found in one or two most likely locations, the cost of the call is also cheaper. Hence the list based scheme outperforms the pointer and the broadcasting schemes under this condition.

Simulations demonstrate that profile information can drastically reduce the number of location updates. In the future, we plan to perform more extensive simulations and also extend our cost model to include the costs of storing profile information. We also plan to simulate a dynamic updating protocol, where the decision whether to update the location or not is based on the recent history of the call to mobility ratio. This is analogous to dynamic replication schemes [25] where the number of replicated copies of data depends upon the read-write pattern on the data item.

B   Query processing

Given that locations will be stored as partitions, exact locations or the paging sequence activity that needs to be carried out will depend upon the query? However, one would like to know a optimal strategy to determine the order of paging.

How hard is it to find a tree corresponding to the optimal strategy?

The answer to this and other questions is provided by the surprising connection with the classification problem in machine learning.

Let $X = \{x_1, \ldots, x_n\}$ be a finite set of objects. Let C be a subset of X. Let $T = \{T_1, \ldots, T_p\}$ be a finite set of tests such that for each of the objects x_i, $T_l(x_i)$ is either true or false. A *classification tree* is a tree built from nodes which correspond to tests $T_1, \ldots, T_p$ and two terminal nodes (POS, NEG). Edges correspond to the results of the test (positive or negative). Each path of the tree ends either in the positive (POS) or negative (NEG) terminal node. The path leads to a positive terminal node if all objects $x \in X$, having tests consistent with the results obtained along the path, belong to C. The path leading to a negative terminal node is defined analogously.

The above definition deals with a binary classification tree. A k-ary tree allows k different results for each test (instead of 2).

The cost of the binary classification tree is the expected length of a path from the root to a terminal node. An optimal (binary) classification tree is the one which minimizes the cost.

Theorem 1: [10]

Given the classification problem as defined above. The following problem is NP-complete: Does there exists a classification tree with the cost below K?

Our problem can be cast as a classification problem in the following way:

The vector $< b_1, \ldots, b_n >$ associated with $x_1, \ldots, x_n$ determines the population of n-ary vectors $l =< l_1, \ldots, l_n >$ such that l_i is in Error(x_i, b_i). This set forms our universe.

Classify these vectors l into two classes: the class of those which satisfy the constraint C $(l_1, \ldots, l_n)$ and the rest. In this way the b_i's act as k-ary "tests" of the k-ary classification problem.

By easy reduction we can show that finding the optimal strategy for an n-ary constraint is also NP-complete. An interesting open problem is to determine the complexity of finding an optimal decision tree when constraint L is syntactically restricted, for instance when it is a conjunction of unary or binary predicates.

The main question however is to what extent does it pay to spend a lot of time on optimizing the search strategy in order to cut down the paging cost? Clearly solving the NP-complete problem in order to save on paging is, in general, not a good idea. It will be acceptable if the paging cost is very large and if incompleteness is not too large, i.e., the cardinalities of $b_1, \ldots, b_n$ are small. The alternative solution is to provide a fast approximate method which would generate sub-optimal query processing strategy. One solution is to apply a similar approach as in machine learning and use greedy approach. A modification of ID3 [21] like algorithm to construct a tree is appropriate. This is a greedy approach which builds the tree by selecting one variable x_k at a time. The variable is selected on the basis of the minimal entropy. Because of the lack of space, greedy algorithm to generate a query evaluation tree is presented in the Appendix.

C Greedy Algorithm for Suboptimal Query Evaluation Strategy

Input: The query $Q(a_1, \ldots, a_n) = \text{Locate}(a_1, l_1) \land Locate(a_2, l_2) \land \ldots Locate(a_n, l_n) \land C(l_1, \ldots, l_n) \land W(a_1, \ldots, x_n)\}$
Relation $Location(a_i, c_i)$ and the error function $Error(a_i, c_i) = b_i$. In other words $b_1, \ldots, b_n$ are partitions where $a_1, \ldots, a_n$ can be found.

Output: The tree which shows possible orderings in which we want to page partitions $b_1, \ldots, b_n$.

We will build a tree with edges labeled by conditions $l_i = d_{ij}$ where d_{ij} is a member of b_i and l_i is location variable (we say that l_i labels this branch). Nodes are either terminal (POS or NEG) or serve as roots for subtrees.

All edges leaving a particular node n are labeled by the same object but differ by a particular value d_{ij} in a condition ($l_i = d_{ij}$). This object is also called a *branching object*.

With each node n of the tree we associate the set path(n) which is a set of all objects l_i labeling edges leading from the root to this node. The set Population(n) denotes a subset of tuples in Y which satisfy conditions along the path leading from the root to n.

Algorithm:
Take a random k-sized subset Y of B = $b_1 \times \ldots \times b_n$ (if the set is B is not very large take the set itself)
N = {root}
Population(root) = Y
While (N <> nil)
Select a node n in N and an object l_i (not in path(n)) as a *branching object* such that the *gain ratio* (an information theoretic measure proposed in [21] and to be defined below) is maximal. Add edges labeled by ($l_i = d_{ij}$) to n and create new nodes n_{ij}. Label n_{ij} as terminal POS node if all tuples in Population(n_{ij}) belong to C. Label it as terminal NEG node if none of the tuples in Population(n_{ij}) satisfies C.
N: = (N - {n}) + all nonterminal nodes n_{ij}
End
Below we define the notion of information gain:
Let the set e = Population(n) contain e_1 objects which satisfy constraint C and e_2 objects which do not. Then the entropy E of Population(n) is given by

$$E = -\sum_{k=1}^{2} \frac{e_k}{e} \log_2 \frac{e_k}{e}$$

If location variable l_i such that $b_i = \{d_{i1}, d_{i2}, \ldots, d_{im}\}$ is used as the branching location variable, it will partition e = Population(n) into $e_{i1}, \ldots, e_{im}$ with e_{ij} containing these tuples from e for which $l_i = d_{ij}$. The expected entropy for the subtree corresponding to subpopulation e_{ij}

is E_{ij}. The expected entropy for all subtrees generated by the branching location variable l_i is then the weighted average

$$E_i = \sum_j \frac{e_i^j}{e} E_j^n$$

The information gain by branching on l_i is therefore

$$\text{gain}(l_i) = E - E_i$$

The information content of the location variable l_i can be expressed as

$$I(l_i) = -\sum_j \frac{e_i^j}{e} \log_2 \frac{e_i^j}{e}$$

The gain ratio for the location variable l_i is then defined to be the ratio

$$\text{gain}(l_i)/I(l_i)$$

Example:
Let $b_1 = \{a, b, c\}$ $b_2 = \{d, e, f\}$ and $b_3 = \{g, h, k\}$ be partitions corresponding to objects a_1, a_2, and a_3 respectively. Assume that the constraint C is satisfied if $l_1 = c$ (no matter what the values for l_1 and l_2 are) and if $l_1 = a$ and $l_2 = d$. Assume also that the constraint is false if $l_1 = b$ (again, no matter what the other values are) or when $l_1 = a$ and $l_2 = d$, or for $(l_1 = a)$, $(l_2 = f)$, $(l_3 = h)$ or finally for $(l_1 = a)$, $(l_2 = f)$, $(l_3 = k)$. For other combination of values the constraint is true.
What is the optimal sequence in which we will broadcast in our search for positions of a_1, a_2, and a_3?
Let us compute information gain for each of the above objects:
For l_1 the gain is the largest (around 0.2), then the next candidate is l_2 and finally l_3. The final tree which determines the broadcasting strategy is shown in Figure 10.
Given a strategy, when we evaluate the query, after each broadcast we have to evaluate the "residual constraint" - that is push the obtained location value "down the constraint tree."

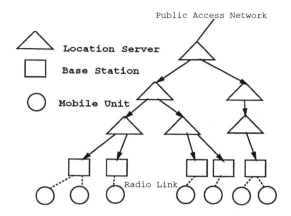

Figure 1: Architecture of Cellular Network

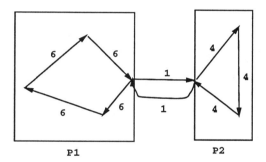

Figure 2: Partitions

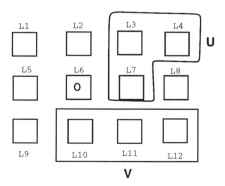

Figure 3: Near By Locations

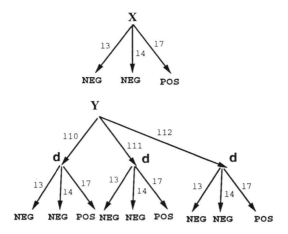

Figure 4: Classification Tree

Example Consider the query to find the number of taxi cabs are in the campus area. Since the database knows only the partitions in which the taxi cabs are located, bounds (between 3 and 6 taxis) can be obtained immediately. However, improving the bounds will require additional paging. An interesting question is by how much can we improve the bound given that we can use only K messages.

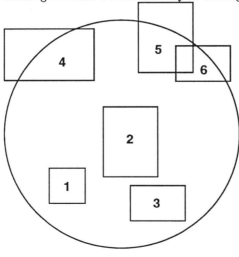

3<= COUNT <= 6

Figure 5: Users in a given area

Example Another example of an aggregate query is to find the distance to the closest doctor. The doctors are known to be in partitions 1, 2, or 3. The closest distance can either be MIN if a doctor is found in partition 1 or MAX if the doctor is found in partition 2. The closet distance will lie within these bounds.

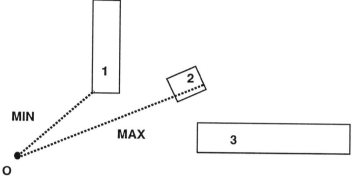

Figure 6: Nearest Object

Example Which highway does User X have to take to get to reach User Y? If user Y is in the circled partition, then this partition does not intersect the alternate route via I94. Hence, the answer is I87.

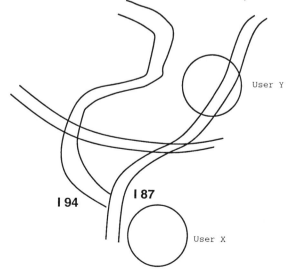

Figure 7: Closet path

106

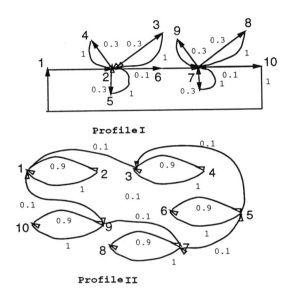

Profile I

Profile II

Figure 8: User Profiles

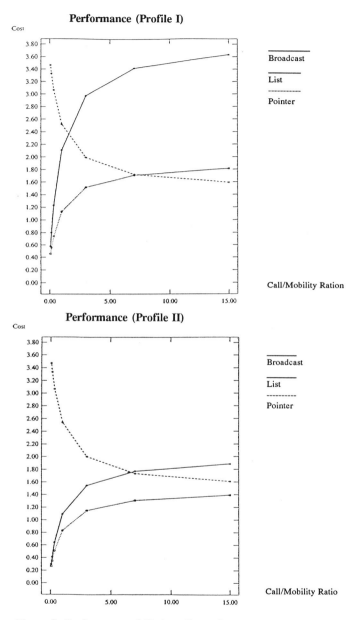

Figure 9: Performance of Various Strategies

108

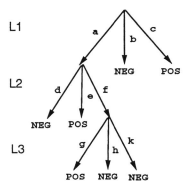

Figure 10: Classification Tree

Design of High-Speed Wireless Links Using Non-Directional Infrared Radiation

J. M. Kahn, J. R. Barry, M. D. Audeh, E. A. Lee and D. G. Messerschmitt

Department of Electrical Engineering and Computer Sciences
University of California, Berkeley, CA 94720

ABSTRACT

We review the advantages of non-directional infrared radiation as a medium to provide wireless indoor access to local-area networks. We describe the major impairments to such non-directional infrared links, which are: small received signal power in the face of potentially intense ambient infrared radiation, optical interference from fluorescent lighting, and intersymbol interference caused by multipath optical propagation. We discuss two communication techniques that can counter these impairments: baseband on-off keying with adaptive decision-feedback equalization and multiple-subcarrier modulation. For the former modulation/equalization technique, we present detailed performance simulations, which show that an adaptive decision-feedback equalizer is effective in mitigating multipath-induced intersymbol interference. Finally, we present an analysis of the signal-to-noise ratio of a receiver for baseband on-off keying. This analysis shows that with a properly designed PIN-FET receiver, it is possible to achieve digital communication at speeds as high as 100 Mb/s.

1.0 Introduction

Non-directional infrared radiation is an attractive transmission medium for wireless indoor access to local-area networks [1]. The most important advantages offered by infrared over radio are the availability of a virtually unlimited, unregulated spectrum, and the fact that infrared radiation does not pass through walls or other opaque barriers. As described here, a single infrared link can operate with a bit rate as high as 100 Mb/s. Since it is possible to operate at least one infrared link in every room of a building without interference, the potential capacity of an infrared-based network is extremely high. We have described in detail the advantages and drawbacks of this transmission medium in a recent publication [2].

Multimedia networking of portable notebook computers is a promising application of wireless infrared communication and has guided our consideration of this technology. The structure of an envisioned local-area network using infrared access is shown in Fig. 1. A high-speed optical fiber backbone would connect information servers to base-station transceivers located in each network cell. Small rooms would contain a single network cell, while rooms larger than about 50 m^2 might contain multiple cells. Portable terminals would display high-resolution graphics and full-motion video, requiring high-speed downlink communications. By contrast, the uplink bit rate might

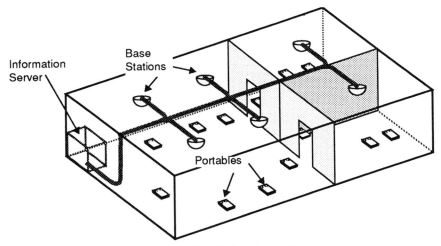

Fig. 1. A local-area network employing wireless infrared access.

be much lower, since the uplink would convey only speech, key strokes and pen movements.

Our studies have focused on the high-speed downlink, since it provides a much greater technical challenge than the uplink. We propose to communicate via non-directional infrared beams that emanate from a ceiling-mounted emitting aperture. Such non-directional infrared beams propagate much like the visible output of lighting fixtures, eliminating the need for alignment between the receiver and transmitter, and providing some resistance against shadowing of the receiver.

We have attempted to identify technologies and communication techniques to allow operation of a non-directional 100 Mb/s link. While there is perhaps no immediate widespread need for such a high bit rate, our work has shown that it is quite possible to achieve a link at this speed. This suggests the immediate feasibility of building non-directional infrared links for portable terminals at somewhat lower bit rates, bypassing the spectral-allocation problems associated with using radio links in a very high-capacity network. We note that there are already commercially available directional infrared links at bit rates of up to 4 Mb/s, incorporated in wireless local-area networks designed to serve fixed desktop terminals.

In this paper, we review the properties of the non-directional infrared channel, describing the impairments to high-speed communication. We describe modulation/demodulation techniques that are appropriate for this channel. Finally, we discuss the design of an optical receiver to allow high-speed communication in the presence of strong ambient infrared radiation.

2.0 The Non-Directional Infrared Channel

The wavelength range between 780 and 830 nm is presently the best choice for non-directional infrared communications. This wavelength band is served by low-cost, highly efficient GaAs-based laser diodes; even more importantly, it is compatible with large-area silicon photodiodes. Achievement of eye safety in the present application will require that the base-station transmitter spread the beam over an extended emitting aperture. If beam spreading is achieved using diffusive reflection or transmission inside the base-station such that the beam's spatial coherence is destroyed, then the resulting radiation is expected to have eye safety equivalent to light emitting diode (LED) radiation. In this case, the maximum permissible exposure (MPE) guidelines [3,4] for LEDs should be applicable. For LED radiation at 900 nm, the MPE is 10 mW/cm^2. We propose to use source powers of the order of 1 W, which would require an emitting-aperture area of at least 100 cm^2 for eye safety. If one is required to limit the transmitter irradiance according to the laser-diode MPE [3,4] of 0.8 mW/cm^2, then the emitting-aperture area must be increased to 1250 cm^2.

Intensity modulation with direct detection is probably the only practical way to use the non-directional optical channel. Coherent optical techniques offers excellent sensitivity and stray-light rejection for intersatellite free-space optical communications [5]. However, the poor spatial coherence of a non-directional beam would lead to a poor photoelectric mixing efficiency, nullifying most or all of the advantage of coherent techniques. Also, the complexity of a coherent receiver is prohibitive in the present cost-sensitive application.

In a non-directional infrared communication link, the existence of multiple paths between transmitter and receiver yields some resistance to shadowing of the receiver. However, it also causes temporal dispersion, which leads to frequency-selective fading. Consideration of multipath optical propagation is, in principle, very similar to that of multipath radio propagation [6], with the important difference that the optical signals are well-confined within the relatively small space of the room of origin. If we could scan a very small power detector (having dimensions smaller than half an optical wavelength) across a room in which a non-directional optical beam undergoes multipath propagation, we would expect to measure a Rician- or Rayleigh-distributed envelope undergoing drastic fluctuations on the scale of half a wavelength. By contrast, an infrared detector of practical size placed at any position in the room would average over all fading conditions, registering an envelope that does not undergo deep fades. To summarize the situation, the non-directional infrared channel with intensity modulation and direct detection can be considered a *multipath baseband channel.*

We have performed detailed numerical simulation of multipath optical propagation, including reflections up to fifth order [7]. Figure 2 presents the results obtained for a particular position in a 5x5 m^2 room having highly reflective walls. Approximately 52% of the received signal power is carried by the line-of-sight (LOS) path. The remaining power is received with a maximum delay of about 50 ns. This temporal dispersion tends to induce intersymbol interference (ISI), impairing high-speed digital communication. For example, Fig. 3 presents an analytical calculation of the optical power penalty induced by the multipath impulse response of Fig. 2(a) when baseband on-off-keyed modulation is used. This analysis assumed that the receiver uses an integrate-and-dump filter, which is the matched filter in the absence of multipath dispersion. As one might expect from inspection of the impulse response, multipath causes substantial impairment to communication at bit rates above approximately 10 Mb/s.

112

Non-directional infrared propagation results in a spreading of the transmitted power over a large area, leading to a small received signal irradiance (power per unit area). The particular example represented in Fig. 2 uses a 1 W source radiating in an ideal Lambertian pattern (intensity proportional to the cosine of the angle from the surface normal); actual diffuse reflectors are well approximated as Lambertian [1]. The received LOS irradiance is 1.23 μW/cm^2, while the total irradiance is 2.38 μW/cm^2. Unfortunately, this received irradiance is much smaller than the potential ambient infrared background light in an indoor environment. Illumination from strong daytime skylight [1] can be as strong as 290 μW/cm^2, even after passage through a 50-nm-wide optical interference filter. Since the passband of an interference filter shifts

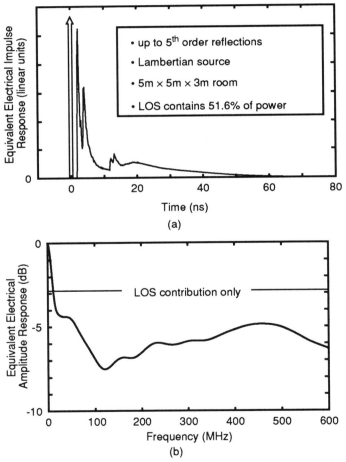

Fig. 2. The simulated impulse response (a) and corresponding amplitude response (b) for a typical non-directional infrared channel, accounting for up to five reflections [7]. The channel input and output are optical intensities. The following parameters were used: room size 5x5 m^2, ceiling height 3 m; reflectivity of the ceiling and each wall 0.8; reflectivity of the floor 0.3; transmitter at center of ceiling pointing straight down with an ideal Lambertian intensity pattern; receiver pointing straight up, located on the floor 1 m from the north wall and 0.5 m from the west wall.

with changing angle of incidence [8], an excessively narrow filter will have a limited field of view. The practical lower limit to filter bandwidth appears to lie in the range between 10-50 nm. As described in Section 4 below, this strong ambient radiation leads to photodetector shot noise that forms the limit to the signal-to-noise ratio of a well-designed receiver. In turn, the need for a large detector, inevitably associated with a high capacitance, tends to limit the receiver bandwidth, thus placing a premium on the electrical bandwidth efficiency of any candidate modulation technique.

Radiation from fluorescent lights contains cyclostationary noise at harmonics of the power-line frequency, extending up to hundreds of kHz. While most of this interference is blocked by a narrow-band optical filter, it is still necessary to eliminate the detected near-dc noise by passage of the received signal through a high-pass electrical filter. This requires either that the transmitted signal not contain significant energy near dc, or that the receiver employ baseline restoration.

3.0 Communication Techniques

There exist several candidate modulation/demodulation techniques for achieving high-speed communication in the face of the impairments described in Section 2.0. Two techniques, baseband on-off-keying with adaptive equalization, and multiple-subcarrier modulation, appear most promising [2].

3.1 Baseband On-Off Keying

Baseband NRZ on-off keying (OOK), as commonly employed in many commercial optical fiber communication systems, offers the advantages of simple modulation/demodulation, efficient use of limited receiver electrical bandwidth, and efficient use of the limited available optical power. To limit the near-dc data content, pre-transmis-

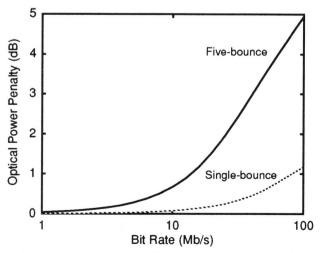

Fig. 3. Optical power penalty due to multipath dispersion as a function of bit rate for unequalized baseband on-off keying. Penalties are shown for impulse responses including contributions from only one bounce, and from up to five bounces.

114

sion coding can be employed, using a technique such as polarity-pulse coding [9]; alternatively, a random NRZ code can be transmitted, and one can implement baseline restoration [10] at the receiver. The major impairment to OOK signaling at 100 Mb/s is ISI induced by multipath propagation, which is expected from analysis to induce an optical power penalty of approximately 4.9 dB (Fig. 3). A standard technique for reduction of ISI is the use of an adaptive decision-feedback equalizer (DFE) [11]. We have performed detailed performance simulations of an OOK system with DFE [12], using the *Ptolemy* simulation under development at U.C. Berkeley.

The block diagram of our simulated system is shown in Fig. 4. Binary data with a rectangular pulse shape are transmitted over the multipath optical channel. This channel is modeled as a tapped delay line that is equivalent to the simulated room multipath impulse response [7] shown in Fig. 2(a). The post-detection filter is either an integrate-and-dump (I&D) or a five-pole Bessel low-pass filter. Since the multipath impulse response is expected to change only on the time scale of tens of milliseconds, it is assumed to be stationary. Thus, everything within the dashed box can be replaced by an equivalent linear time-invariant system sampled every T/2 seconds, reducing simulation run time. This necessitates passing the additive white Gaussian noise through a separate but identical post-detection filter. The filtered data and noise then pass to a DFE that has a T/2-spaced forward filter and a T-spaced reverse filter. The fractionally spaced forward filter is advantageous because it allows the receiver, for the same number of taps, to be less sensitive to timing errors and still function as well as a T-spaced filter [11]. Both the forward and reverse filters are adaptive, with tap updating according to the LMS algorithm.

Figure 5 illustrates the bit-error-rate (BER) vs. average optical power performance of the simulated system. The reference point is 10^{-6} BER for the ideal channel (no multipath) with an I&D filter, as calculated from the Q function. Under these ideal conditions, the simulated points match the theory well. When multipath propagation is included and an I&D filter is used, the simulated system incurs an optical power penalty of about 4.6 dB, in good agreement with the analytical results shown in Fig. 3. This agreement provides assurance of the accuracy of our simulations when they are extended to include effects that are not amenable to analysis. With multipath and an I&D filter, the best simulation results are obtained with a DFE having 3 forward taps and 5 reverse taps; this DFE is able to reduce the multipath ISI penalty to 2.0 dB.

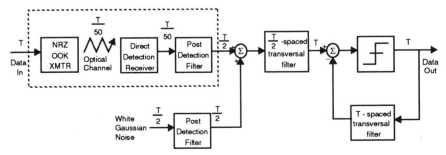

Fig. 4. Block diagram of simulated infrared link using baseband OOK with an adaptive DFE. Shown after each block is the sampling interval at that point in the simulation. The two post-detection filters are identical.

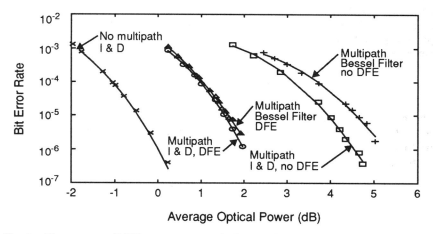

Fig. 5. Bit-error-rate (BER) vs. average optical power for baseband OOK at 100 Mb/s. The reference point (0 dB) is where the ideal curve crosses 10^{-6} BER. This ideal curve represents a system with no multipath propagation and with an integrate-and-dump (matched) filter, and is obtained by evaluation of the Q function. The various symbols represent data points obtained from simulation, and t he other four curves represent least-squares fits to data from the simulation results.

It is difficult to implement an I&D post-detection filter at high speeds, and a five-pole Bessel low-pass filter represents a more practical choice. By simulation it was determined that with the DFE, the optimal 3-dB filter bandwidth is 45 MHz. BER results using a Bessel filter of this bandwidth are shown in Fig. 5, both with and without the DFE (again, the DFE has 3 forward taps and 5 rese taps). Our data indicate that the Bessel filter-DFE combination performs almost as well as the I&D-DFE combination, with a multipath ISI penalty of 2.1 dB. In practice, such a DFE would need to adaptively adjust as the receiver is moved. Because the impulse response is expected to change only on the time scale of tens of milliseconds (i.e., millions of bits), adaptive DFE should be a feasible technique for overcoming multipath optical propagation. We are presently extending these simulations to investigate the effect of a non-stationary multipath response on the phase of a recovered clock, the ability of the DFE to compensate for these timing errors, and the bit accuracy required in a practical implementation of the DFE.

3.2 Multiple-Subcarrier Modulation

If individual mobile users need to receive the entire 100-Mb/s bit stream, then OOK with adaptive DFE is an attractive technique. However, if the 100-Mb/s stream is used to convey multiplexed lower-bit-rate data to several users, OOK modulation presents a drawback: it requires that each receiver detect the aggregate 100-Mb/s stream and perform digital demultiplexing to obtain the desired data. This potential drawback of OOK is naturally overcome by using an alternative modulation technique, multiple-subcarrier modulation [13]. Because it opens up the possibility of multi-level (M-ary) modulation and frequency-division multiplexing while retaining the simplicity of intensity modulation and direct detection, this technique has received widespread consideration for optical-fiber transmission of video signals [13].

116

The principle behind multiple-subcarrier modulation is illustrated in Fig. 6. As a simple example, Fig 6(b) shows the transmitted and received signal waveforms for a single binary phase-shift-keyed (BPSK) subcarrier. The BPSK-modulated carrier u(t) can be described by:

$$u(t) = cos(2\pi f_1 t) \sum_k a_k p(t - kT) \qquad (1)$$

where $a_k \in \{-1, 1\}$ represents the data, f_1 is the subcarrier frequency (say of the order of 50 MHz), p(t) is the baseband pulse shape and T is the symbol interval. The electrical signal u(t) can be transmitted on an optical carrier ($f \sim 10^{14}$ Hz) by modulating the intensity of the light with u(t):

$$p(t) = p_{max} \frac{[1 + u(t)]}{2} \qquad (2)$$

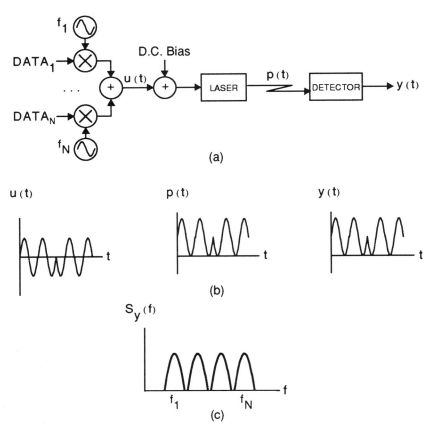

Fig. 6.   Optical subcarrier modulation. (a) Block diagram of a multiple-subcarrier system. (b) Signal waveforms in single-subcarrier BPSK system, showing laser drive signal u(t), transmitted optical intensity p(t), and detected photo current y(t). (c) Power spectrum of y(t) in a multiple-subcarrier system.

where p_{max} is the peak power output of the optical source. Since a photodetector responds only to the intensity of the incident light, the electrical signal y(t) at the receiver is simply u(t) corrupted by the noise and multipath distortion of the optical channel. Using standard radio frequency techniques, this electrical signal y(t) is demodulated to yield the data a_k. As indicated in Fig. 6(c), several carriers that convey independent data can be frequency-division multiplexed to form the signal u(t). The individual data streams can be recovered from y(t) by using multiple bandpass demodulators.

Because each of the multiple subcarriers is detected in a frequency range away from DC, this technique is naturally immune to interference from fluorescent lights. To understand the impact of multipath propagation on multiple-subcarrier transmission, it is useful to consider the amplitude response of a typical non-directional optical channel, as illustrated in Fig. 2(b). Suppose that the transmitted signal u(t) consists of several subcarriers, each having a relatively low signaling rate of the order of 10 Mbaud, spaced across the band below 100 MHz. Then in the detector output y(t), each subcarrier will be attenuated differently, according to the channel amplitude response of Fig. 2(b). However, the attenuation undergone by each subcarrier will be approximately constant across the relatively narrow frequency band it occupies, so that each subcarrier will suffer little distortion due to multipath propagation. The relative attenuation of one subcarrier relative to another is corrected easily through the use of an independent automatic gain control circuit in each bandpass demodulator.

The received signal spectrum of Fig. 6(c) suggests an additional potential advantage of multiple-subcarrier modulation, namely, the ease of multiplexing and demultiplexing the data required by multiple users of the optical channel. If the 100 Mb/s aggregate link capacity is to be allocated to several concurrent users, then each user's bandpass demodulator can be set to selectively receive only the data intended for that user. This avoids the need for each user to possess electronics capable of demodulating and demultiplexing the entire link bit rate. Such a scheme would also support multiple terminal types capable of receiving different bit rates, and even simultaneous analog transmission.

The main prices paid for the advantages of subcarrier modulation are a potential reduction in link power budget and a possible increase in required receiver electrical bandwidth. Considering a peak-power-limited optical source, subcarrier techniques suffer at least a 3-dB loss of optical power efficiency, as compared to baseband OOK. Multiple-subcarrier modulation suffers an additional loss of power efficiency because the channel has greater attenuation at higher subcarrier frequencies [see Fig. 2(b)]. It is possible to avoid additional degradation due to source nonlinearities [13] though the use of a separate transmitting laser for each subcarrier. Although multiple-subcarrier modulation suffers a loss of peak-power efficiency, the low bit rate of each individual channel might permit the implementation of powerful error-correction coding capable of achieving coding gains in excess of the loss of power efficiency.

4.0 Receiver Design

Even when high-power sources and narrow-band receiver filters are used, non-directional infrared propagation can result in a signal irradiance at the receiver that is of the order of 100 times smaller than the ambient infrared irradiance, particularly when strong skylight is admitted to a room [1]. This situation contrasts strongly with optical fiber communications, where virtually no background light is received. In non-directional infrared systems, the strong ambient light induces shot noise in the receiver; forming the limiting factor in the SNR of a properly designed receiver. Because of this strong background irradiance, design of a non-directional infrared receiver requires a strategy very different from that commonly employed in optical fiber communications. In this section, we outline that strategy and show that it is quite feasible to achieve a high receiver SNR using easily available, inexpensive optical and electrical technologies. We confine the discussion to the case of a baseband receiver for OOK; the extension to a passband subcarrier receiver is straightforward.

4.1 Selection of Receiver Components and Preamplifier Type

In order to reduce the ambient infrared irradiance to the smallest possible value, it is desirable to employ a narrow-band optical filter at the receiver. This filter should have a bandpass as narrow as possible, while passing the transmitted signal with little attenuation. For optical bandpass interference filters, the wavelength of peak transmittance shifts with varying angle of incidence [8], implying that an excessively narrow filter will result in a reduced receiver field of view. In section 4.3, we will provide a figure of merit to aid in the choice of an optimum optical bandpass filter.

There are two candidate types of photodetectors for non-directional infrared receivers: positive-intrinsic-negative photodiodes (PINs) and avalanche photodiodes (APDs) [14]. For the wavelength band between 780 and 830 nm, excellent large-area silicon devices of both types are available. A PIN detector consists of a reverse-biased diode that is capable of converting received optical power to current with a conversion efficiency called the responsivity R, which is of the order of 0.5 A/W for detectors in this wavelength band. An APD detector consists of a specially designed PIN detector to which one applies a high reverse bias (of the order of 200 V), bringing the device to the threshold of avalanche breakdown, and allowing it to multiply the photo-induced current by a current gain M, which typically lies in the range between 1 and 100. The internal electrical gain of the APD reduces the relative impact of preamplifier circuit noise on the receiver SNR performance. Unfortunately, the gain of an APD is itself noisy, and the shot-noise-to-signal ratio of an APD is increased over that of a PIN detector by a gain-dependent excess-noise factor F(M). In optical fiber receivers [14], where the shot noise arises only from the weak signal, this increased shot noise is outweighed by the decreased impact of circuit noise, and an APD provides a dramatic SNR enhancement. By contrast, shot noise in the present application is much stronger relative to the signal, and the APD noise enhancement can actually degrade the overall receiver SNR.

The analysis of Section 4.3 will demonstrate that in a well-designed non-directional infrared receiver, achievement of a high receiver SNR requires a large detector area. A large physical detector size will result in increased cost, as well as increased receiver input capacitance. It is therefore highly desirable to increase the effective detector light collection area through the use of a hemispherical condensing element [15]. When placed in front of the flat photodetector, a hemispherical lens having refractive index n can provide an optical gain G of up to n^2, which can be more than 3 dB for common optical materials.

The present application obviously demands a high receiver dynamic range, implying that it is desirable to use a transimpedance-type preamplifier [14], in which negative feedback is used to compensate for the dominant pole introduced by the photodetector capacitance and the resistance used to convert the photocurrent to a voltage.

Our notation for parameters used in the analysis, as well as typical values assumed or derived in our example, are presented in Table 1.

Table 1: Receiver design variables and values assumed or derived in example.

Variable	Definition	Value
A	Photodetector area	1.34 cm^2
A_v	Transimpedance amplifier open-loop voltage gain	11.3 V/V
B	Bit rate	100 Mb/s
c_d	Photodetector capacitance per unit area	35 pF/cm^2
C_d	Photodetector capacitance	46.9 pF
C_g	FET gate capacitance ($C_{gs} + C_{gd}$)	205 fF
f_T	FET unity-gain cutoff frequency	44 GHz
f_c	FET 1/f-noise corner frequency	10 MHz
F_{mp}	Electrical SNR penalty due to multipath interference	1 (0 dB)
F(M)	APD photodetector excess-noise factor	1 (PIN used)
g_m	FET transconductance	57 mS
G	Receiver optical gain	2.0 W/W
i_{sig}	Received peak signal photocurrent	1.31 μA
i_{bg}	Received background photocurrent	411 μA
i^2_{FET}	Input-referred noise current due to FET channel noise and 1/f noise	$2.2 \times 10^{-15} \text{ A}^2$
i^2_{shot}	Input-referred noise current due to shot noise	$7.4 \times 10^{-15} \text{ A}^2$
i^2_{res}	Input-referred noise current due to feedback resistor	$2.2 \times 10^{-15} \text{ A}^2$
i^2_{tot}	Total input-referred noise current	$11.8 \times 10^{-15} \text{ A}^2$

Table 1: Receiver design variables and values assumed or derived in example.

Variable	Definition	Value
I_{bg}	Received background irradiance (after optical bandpass filter)	290 μW/cm^2
I_{sig}	Received signal irradiance (after optical bandpass filter)	0.92 μW/cm^2
I_f	Noise bandwidth factor for FET 1/f noise	0.184
I_2	Noise bandwidth factor for shot and resistor noises	0.562
I_3	Noise bandwidth factor for FET channel noise	0.0868
M	APD photodetector current gain	1 (PIN used)
r_{FET}	Ratio of noise variances i^2_{FET} / i^2_{shot}	0.30
r_{res}	Ratio of noise variances i^2_{res} / i^2_{shot}	0.30
R	Photodector responsivity for M=1 ($\eta e\lambda/hc$)	0.53 A/W
R_L	Feedback resistance	417 Ω
$SNR_{tot,req}$	Peak receiver electrical signal-to-total-noise ratio required, no multipath	144 (21.6 dB)
Γ	FET channel noise factor	0.82
η	Photodector quantum efficiency	0.79
λ	Optical wavelength	830 nm

4.2 Analysis of the Receiver Signal-to-Noise Ratio

Our analysis follows the method of previous treatments [14,16,17,18] of optical-fiber receivers, and makes reference to the receiver structure illustrated in Fig. 7. The received optical pulses are assumed to be rectangular in time; the effect of multipath optical propagation will be introduced later as a receiver SNR penalty. The transimpedance amplifier consists of a common-source FET amplifier followed by a voltage gain A_2; the open-loop voltage gain of the two cascaded stages is $A_v = g_m R_D A_2$. The transimpedance amplifier output is equalized to a 100%-excess-bandwidth raised-cosine pulse, allowing our analysis to neglect the effect of intersymbol interference. The equalizer analyzed can be well approximated in practice by a properly chosen low-pass filter [14]. The present analysis neglects photodetector dark current and FET gate leakage current, as well as any noise originating in the amplifier A_2 or in following stages. It also neglects any sources of receiver input capacitance other than the photodetector or the FET.

Let I_{sig} and I_{bg} respectively denote the received signal and background irradiances after the optical bandpass filter. Assuming an optical gain G, and a detector area having A, responsivity R, and current gain M (a PIN detector corresponds to M=1) the detected signal and background photocurrents are given by:

$$i_{sig} = I_{sig} \, G \, A \, R \, M \tag{3}$$

$$i_{bg} = I_{bg} \, G \, A \, R \, M. \tag{4}$$

We now consider the sources of receiver noise. When no ISI is present, the receiver bit-error-rate performance is completely determined by the SNR present at the receiver output when it is sampled by the decision circuit. To simplify application of our analysis, we define input-referred noise-current variances (i^2_{shot}, i^2_{res}, i^2_{FET}) whose sum is:

$$i^2_{tot} = i^2_{shot} + i^2_{res} + i^2_{FET}. \tag{5}$$

The various contributions to i^2_{tot} contain noise bandwidth factors (I_2, I_3, I_f) taking into account the transmitted and equalized pulse shapes. Using this notation, the total receiver output SNR at the sampling instant is given simply as:

$$SNR_{tot} = \frac{i^2_{sig}}{i^2_{tot}}. \tag{6}$$

The input-referred shot noise is given by:

$$i^2_{shot} = 2 \, q \, M^2 \, F(M) \, I_{bg} \, G \, A \, R \, I_2 \, B, \tag{7}$$

where B is the transmitted bit rate. The input-referred resistor noise is:

$$i^2_{res} = \frac{4kT}{R_L} \, I_2 \, B, \tag{8}$$

where R_L is the feedback resistance. The final noise term is the input-referred FET noise:

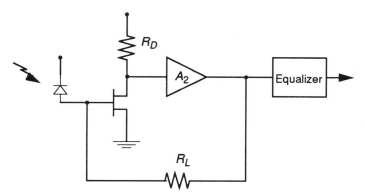

Fig. 7.   Simplified schematic diagram of the transimpedance receiver whose noise performance is analyzed here. The equalizer converts the rectangular received pulses to 100%-excess-bandwidth raised-cosine pulses.

$$i^2_{FET} = \frac{4kT\Gamma}{g_m} \left[2\pi B\,(c_d A + C_g)\right]^2 (I_f f_c + I_3 B), \qquad (9)$$

where Γ is the FET channel noise factor, g_m is the FET transconductance, c_d is the detector capacitance per unit area, C_g is the FET gate capacitance (equal to $C_{gs} + C_{gd}$), and f_c is the corner frequency for FET 1/f noise. We note that for a given FET technology and gate length, C_g and g_m are not independent, but are related by:

$$f_T = \frac{g_m}{2\pi C_g}, \qquad (10)$$

where f_T, the short-circuit common-source current-gain-bandwidth product, is approximately constant. It is useful to display (9) in an alternative form:

$$i^2_{FET} = \frac{8\pi kT\Gamma}{f_T} \frac{(c_d A + C_g)^2}{C_g} (I_f f_c B^2 + I_3 B^3), \qquad (11)$$

For a given f_T, (11) includes the inverse dependence of i^2_{FET} on g_m via the C_g in its denominator.

It is instructive to consider the functional dependence of the various noise terms on the system and circuit parameters. The shot-noise variance is proportional to the dc background-induced photocurrent (4), and is also proportional to the factor M F(M). Temporarily neglecting the resistor and FET noises, the signal-to-shot-noise ratio is:

$$\frac{i^2_{sig}}{i^2_{shot}} = \frac{(I_{sig}GARM)^2}{2qM^2F(M)I_{bg}GARI_2B} = \frac{I^2_{sig}}{I_{bg}} \frac{GAR}{2qF(M)I_2B}. \qquad (12)$$

Expression (12) shows that the signal-to-shot-noise ratio is directly proportional to the square of the signal photocurrent, while it is inversely proportional to the first power of the background photocurrent. The signal-to-shot-noise ratio is inversely proportional to F(M), the APD excess-noise factor, while it is independent of the APD current gain M. If we assume that a receiver is substantially shot-noise limited and that all other parameters are equal, this dependence indicates that an APD-based receiver has a lower SNR than a PIN-based receiver [M = F(M) = 1].

Equation (8) indicates that in order to reduce the resistor noise to an acceptable level, the value of R_L must be chosen to be sufficiently large. Expression (11) shows that the amplifier noise increases quadratically with the detector area A. This implies that as we increase A to improve the signal-to-shot-noise ratio (holding all other parameters fixed), we will reach an area beyond which the total SNR does not increase further, because increased amplifier noise will nullify the improvement. In order to minimize amplifier noise, one should choose an FET having a large value of f_T and a small value of f_c. Given that type of FET, the noise (11) is minimized if one chooses the FET gate width such that $C_g = c_d A$, i.e., the FET gate capacitance should equal the photodetector capacitance.

In general, the design of a receiver proceeds as follows. We begin by assuming that, in the absence of multipath propagation, the receiver is required to achieve a total peak

SNR of $SNR_{tot,req}$ [a bit error rate of 10^{-9} in an OOK system implies $SNR_{tot,\ req} = 144$ (21.6 dB)]. Assuming that multipath interference induces an electrical SNR penalty F_{mp}, we require that the signal and total noise variance, which we calculated above ignoring multipath, satisfy:

$$\frac{i^2_{sig}}{i^2_{tot}} = F_{mp}\ SNR_{tot,req}. \qquad (13)$$

We then choose the system and circuit parameters to satisfy (13). We should note two important practical restrictions on choice of circuit parameters. The feedback resistance R_L cannot be made arbitrarily large because this will move the dominant pole toward dc, requiring an unattainable open-loop gain to provide sufficient receiver bandwidth. Also, it is generally not desirable to choose the FET gate capacitance C_g equal to the detector capacitance c_dA, since this would imply unrealistically large gate width and FET drain current, leading to unacceptable power consumption.

4.3 Simplified Design Procedure for PIN-Based Receivers

We now describe a simple procedure for the design of PIN-based receivers. While the present procedure could in principle be used for design of APD-based receivers, it hides the detailed interplay of the APD parameters A, M and F(M) with the preamplifier circuit parameters. In any case, it is desirable to avoid the high bias voltage required by an APD. Fortunately, as shown by the design example of Section 4.4, it is possible to achieve a very high SNR using a PIN-based receiver.

As previously mentioned, there are practical limitations on the reduction of resistor and FET noise. Rather than performing a multivariate optimization subject to bandwidth and power-consumption constraints, we simply accept that resistor noise and FET noise are present, and assume that their variances lie in some fixed ratios to the shot noise:

$$i^2_{res} = r_{res}\ i^2_{shot} \qquad (14)$$

$$i^2_{FET} = r_{FET}\ i^2_{shot}, \qquad (15)$$

so that the total noise is related to the shot noise by:

$$i^2_{tot} = (1 + r_{res} + r_{FET})\ i^2_{shot}. \qquad (16)$$

We identify the quantity $(1 + r_{res} + r_{FET})$ as the SNR penalty arising from resistor and FET noise. Using (16), condition (13) becomes:

$$\frac{i^2_{sig}}{i^2_{shot}} = (1 + r_{res} + r_{FET})\ F_{mp}\ SNR_{tot,req}. \qquad (17)$$

Using (12) to replace the left-hand side of (17), and solving for the photodetector area A, we find an expression for the required detector area:

$$A_{req} = \frac{I_{bg}}{I^2_{sig}} \frac{2qF(M)I_2B}{GR} (1 + r_{res} + r_{FET}) \; F_{mp} \, SNR_{tot,req.} \qquad (18)$$

Next, knowing the required detector area, we use (7) to calculate the shot noise variance. Using (14) and (15), we calculate the required resistor and FET noise variances. Finally, we adjust the resistor and FET parameters so that these variances are not exceeded.

It is desirable to minimize the required detector area A_{req}. Assuming that the resistor noise, FET noise and multipath penalties are *fixed*, A_{req} is minimized by: utilizing as high a source power as possible, choosing a detector responsivity R as large as possible, choosing a PIN detector [which has F(M) = 1], and by choosing a combination of hemispherical lens and optical bandpass filter to maximize the ratio $I^2_{sig} \, G \, / \, I_{bg}$.

If one designs the receiver allowing a power margin for detector shadowing, the total detector area may be smaller if one uses several small detectors, as opposed to a single large detector. For example, suppose that one wants to design a single-detector receiver with a 6-dB optical power margin to guard against shadowing or tilting of the receiver. Considering a fourfold reduction of I_{sig}, (18) shows that we require a sixteen-fold increase of A (if I_{bg} is reduced also, the required increase will be somewhat less). However, it is probably a better strategy not to increase the detector area by such a large amount, but rather to employ two detectors whose difference in position and tilt guards against simultaneous loss of signal power in both. The two photocurrents can be preamplified separately, and one can process the resulting electrical signals using either a signal-combining or switched-diversity technique [15].

4.4 Example of PIN-Based Receiver Design

In this section we present an example of PIN-based receiver design, showing that with a realistic choice of system and circuit parameters, it is possible to achieve a SNR permitting reliable communication at 100 Mb/s using OOK modulation. The parameters assumed or derived here are summarized in Table 1.

We assume a room configuration [7] identical to that used to calculate the multipath responses illustrated in Fig. 2. In this room, which has highly reflective walls, multipath propagation leads to an increase of received power, but this is accompanied by an optical power penalty due to multipath-induced ISI. Specifically, assuming a transmitted power of 1 W and no optical filter losses, the received line-of-sight (LOS) signal irradiance is 1.23 $\mu W/cm^2$ and the total received irradiance is 2.38 $\mu W/cm^2$, so that multipath provides a 2.9-dB increase in received irradiance. By comparison, when a DFE is used to correct multipath ISI, simulations show that the ISI-induced optical power penalty is only 2.1 dB [12]. In other words, multipath is actually expected to induce a *net improvement* in system power budget. If we instead consider that the room has black walls, then there is no multipath ISI penalty ($F_{mp} = 1$), but the total received irradiance is only 1.23 $\mu W/cm^2$. The design example of this section conservatively assumes this latter case. Allowing for an optical filter transmission of 75%, the net signal irradiance is $I_{sig} = 0.92 \; \mu W/cm^2$.

We assume strong skylight and an optical filter bandwidth of 50 nm, so that the background irradiance [1] is 290 $\mu W/cm^2$. The receiver uses a hemispherical lens having a refractive index of 1.7 (maximum gain of 2.89), which conservatively achieves a gain G = 2.0. The resistor- and FET-noise variances are each assumed to be 30% of the shot-noise variance, implying a total electrical SNR penalty of 2.0 dB. Assuming a PIN detector having responsivity of 0.53 A/W, we obtain a required detector area of A_{req} = 1.34 cm^2. The total background-induced photocurrent of 411 μA leads to a shot-noise variance of 7.4 x 10 $^{-15}$ A^2. The total detector capacitance is C_d = 46.9 pF, assuming a capacitance per unit area of c_d = 35 pF/cm^2.

In order to achieve the required FET-noise variance of i^2_{FET} = 0.3 i^2_{shot}, we choose FET parameters typical of low-noise microwave high-electron-mobility transistors (HEMTs): f_T = 44 GHz, Γ = 0.82, f_c = 10 MHz. Using (11), we see that the gate capacitance must be at least C_g = 205 fF to achieve sufficiently low FET noise. Such an FET could operate with a drain current as small as about 10 mA, for a power consumption of approximately 20 mW.

Finally, achieving the required resistor-noise variance of i^2_{res} = 0.3 i^2_{shot} would necessitate a feedback resistance of at least R_L = 589 Ω. In view of the total receiver input capacitance of C_d + C_g = 47.1 pF, the receiver open-loop bandwidth is $[2\pi R_L(C_d + C_g)]^{-1}$ = 8.1 MHz. Assuming that the receiver 3-dB bandwidth must equal the bit rate, 100 MHz, this open-loop bandwidth would require that the transimpedance amplifier have an open-loop gain of at least 11.3 V/V, which should be achieved easily using a modest number of stages.

5.0 Conclusions

Non-directional infrared radiation can provide an indoor communication medium that is unconstrained by regulatory bandwidth allocations and is free from crosstalk between network cells in adjacent rooms. The primary impairments to non-directional infrared communication are a small received irradiance, a large ambient infrared irradiance, interference from fluorescent lighting, and multipath optical propagation. This paper has described two communication techniques that appear capable of overcoming these limitations: baseband on-off-keying (OOK) and multiple-subcarrier modulation. We have described detailed simulations of an OOK system, which indicate that multipath-induced degradation can be reduced substantially using an adaptive decision-feedback equalizer. Furthermore, this paper presented an analysis of the signal-to-noise ratio in a baseband receiver for OOK. Assuming realistic system and circuit parameters, we found that it is feasible to achieve reliable digital communication at 100 Mb/s using a PIN/FET receiver.

6.0 Acknowledgments

This research has been supported by National Science Foundation PYI Award ECS-9157089, a National Science Foundation Graduate Fellowship, Digital Equipment Corporation, International Business Machines Corporation, and the State of California MICRO Program. One of us (J.M.K) is pleased to acknowledge helpful discussions with Professor P.R. Gray.

References

1. F. R. Gfeller and U. H. Bapst, "Wireless In-House Data Communication via Diffuse Infrared Radiation" *Proceedings of the IEEE* **67**, pp. 1474-1486 (1979).

2. J.R. Barry, J.M. Kahn, E.A. Lee and D.G. Messerschmitt, "High-Speed Non-Directive Optical Communication for Wireless Networks", *IEEE Network Magazine*, pp. 44-54, November, 1991.

3. "Threshold Limit Values and Biological Exposure Indices for 1988-1989", *ACGIH*, Cincinnati, OH, 1988.

4. "American National Standard for the Safe Use of Lasers", *ANSI / Z136.1*, 1986.

5. H.K. Phillipp, A.L. Scholtz, E. Bone, and W. R. Leeb, "Costas Loop Experiments for a 10.6 μm Communications Receiver", *IEEE Trans. on Commun.* **COM-31,** pp. 1000-1002, 1983.

6. J.G.Proakis, *Digital Communications Second Ed.*, Mc Graw-Hill, *1989.*

7. J. R. Barry, J. M. Kahn, E. A. Lee and D. G. Messerschmitt, "Simulation of Multipath Impulse Response for Wireless Optical Channels", to be published in *IEEE J. Sel. Areas Comm.*

8. Optical Coating Laboratory Incorporated, 1990 Product Catalog.

9. K.W. Cattermole, *Principles of Pulse Code Modulation*, London: Iliffe Books, 1969.

10. B. Enning and R. S. Vodhanel, "Adaptive Quantized Feedback Equalization for FSK Heterodyne Transmission Experiments at 150 Mb/s and 1 Gb/s", *Electron. Lett.*, **24**, pp. 397-399, 1988.

11. E.A. Lee and D.G. Messerschmitt, *Digital Communication*, Kluwer Academic Press, 1988.

12. M. D. Audeh and J. M. Kahn, "Simulation and Evaluation of Baseband OOK Modulation for Indoor Infrared LANs at 100 Mb/s", *Proc. of IEEE Int. Conf. on Sel. Topics in Wireless Comm.*, June 25-26, 1992, Vancouver, BC.

13. T. E. Darcie, "Subcarrier Multiplexing for Lightwave Networks and Video Distribution Systems", *IEEE J. Sel. Areas in Comm.* **7**, pp. 1240-1248, 1990.

14. B.L. Kasper, "Receiver Design", pp. 689-723 in *Optical Fiber Telecommunications II*, S.E. Miller and I.P. Kaminow, Eds., Academic Press, 1988.

15. M. D. Kotzin, "Short-Range Communications Using Diffusely Scattered Infrared Radiation", PhD Thesis, Northwestern University, 1981.

16. S.D. Personick, "Receiver Design for Digital Fiber Optic Communications Systems, I and II", *Bell Syst. Tech. J.* **52**, pp. 843-886, 1973.

17. T.V. Muoi, "Receiver Design for High-Speed Optical-Fiber Systems", *IEEE J. Lightwave Technol.* **LT-2,** pp. 243-267, 1984.

18. T.V. Muoi, "Optical Receivers" pp. 441-472 in *Optoelectronic Technology and Lightwave Communications Systems*, C. Lin, Ed., Van Nostrand Reinhold, 1989.

On the Capacity of Time-Space Switched Cellular Radio Link Systems for Metropolitan Area Networks

Jens Zander
Radio Communication Systems
Royal Institute of Technology
Electrum 207
S-164 40 STOCKHOLM-KISTA, SWEDEN

and

Karl-Axel Åhl
Time Space Radio AB
IDEON
S-223 70 LUND, SWEDEN

Abstract

Using multipoint-radio link systems constitutes a viable alternative to copper and optical fibres for high capacity local distribution of telephone and data traffic. In the paper a novel architecture for a fixed cellular radio system is proposed. The proposed system uses electronically steerable antennas to combine time and "space" multiplexing. Assuming a high user-density urban environment, the performance of the system is estimated for some simple propagation and interference models. Results indicate that, due to the time-space multiplexing technique, the performance exceeds the performance of conventional multipoint systems by an order of magnitude. Due to the unique features of the system, several central stations may be operated at the same site at the same frequency to boost the capacity even more.

1. Introduction

The idea to use point-to-multipoint (P-MP) radio link systems in rural area local loops has been proposed many years ago. In, fact several, systems, both analog as well as digital are commercially available today[1,2]. These systems are very similar to cellular radio system in the sense that a central station using an omnidirectional (or sectorized) antenna to serves a large number of peripheral stations. The peripheral stations use simple directional antennas pointed at the central station. In analog systems FDMA-techniques are normally used, whereas TDM/TDMA-techniques are common in digital system. Using directional antennas in the periferal stations allows for good range also at very high frequencies. Due to the low cost of installation compared to rural copper wiring costs, these radio systems are providing an economically viable solution in many rural environments. P-MP radio link systems have lately also been proposed for the metropolitan telecommunication market. Traditional telephone operators now face a new challenge by the introduction of wireless local loop solutions[3]. However, so far, existing systems have failed to meet the demands for high capacity and spectral efficiency required in this type of application.

An interesting advantage, beside that of the improvement of the link-budget, is that directional antennas will suppress interference from surrounding transmitters in the same frequency band. In a similar manner, a transmitter fitted with a highly directional antenna will cause little interference to neighboring receivers. Thus, a way to increase the performance of conventional P-MP systems would be to use directional antennas also at the central station site. This has been proposed in the context of cellular mobile communications [4]. Straigth-forward implemention in a fixed P-MP system, would require the use of several fixed antennas or a multiple-fixed spot beam for each of the periferal station.

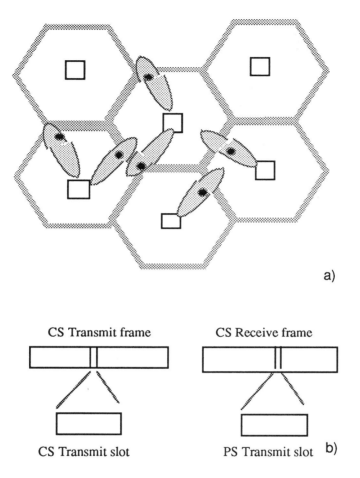

Fig 1. a) System configuration b) Time slot structure.

This concept has been proposed in several satellite systems. In this paper, however, we will consider a TDMA radio link system where the central station uses an electronically steerable, scanning antenna. The system operates in a semi-duplex mode where different time slots are used for communication with the various peripherals. In each time-slot the the central-station redirects its antenna to point to the peripheral in question. Thereby, the interference in other directions is greatly reduced. In addition, the scanning approach offers the flexibility and easy adaption to new or relocated users. This switching concept has been coined Time-Division Beam

Scanning or *Time-space* multiplexing. Systems of this type have been investigated for satellite use [5,6], but also ground radio link system for metropolitan area use have been proposed. In [7], Acampora et al. present a multiple scanning beam arrangement where the central station may transmit in several directions simultaneously. Also the architecture and transmission quality aspects of such a system are discussed in [7]. In [8] a multicell, area covering system is proposed. The aim of this paper is to estimate the interference-limited performance of such a multicell Time-Space Radio (TSR) link system in an urban, high user-density environment. Further, we will compare the results with the performance of some existing system concepts.

2. Cellular TSR Systems

In the following we will investigate a cellular TSR system designed for hexagonal cells. Our system is a single frequency time-duplex TDM-system (fig 1a). The transmit-receive timing of the central station is described in fig 1b). The time-axis is divided into alternating send and receive time frames of roughy equal size. The central station will transmit during its transmit frame and receive during its receive frame. The frames are subdivided into slots, each capable of carrying one data packet. Each active peripheral station will be assigned one or several slots in the receive and transmit frame according to a *Slot Assignment Algorithm* (SAA). The number of slots assigned to each peripheral station in a frame will reflect its present demand for communication capacity.

Typical frame-sizes may be in the order of milliseconds, whereas the slots have a duration in microsecond range. The propagation delays in cells of reasonable size (cell radius ≈ 1-5 km) are considerably smaller than the frame size, but not neglectable compared to the slot size. Propagation delay compensation has to be performed within each cell in order to avoid slot overlap at the central station receiver. We may achieve this by letting the peripherals start their transmission early "before" each slot. This should cause no problem in a fixed system with readily measurable propagation delays. It is also resonable to assume that central stations in different cells are in (rough) frame-synchronism, but not slot-synchronized.

The central stations in our proposed system are placed roughly at the center of the cells. Each central station uses K steerable "pencil-beam"-type antennas, each covering a (K/360)°-sector, yielding a total 360° coverage (Fig 1a). The central station antennas are redirected for each transmission slot. The peripheral stations use fixed directional antennas pointed at their designated central station. Each sector is at the central station site handled by an individual *(sector) sub-system*, allowing for simultaneous transmissions in all K sectors. The sub-systems in a cell are assumed to be in frame synchronism to avoid that a transmission is in progress in one sector sub-system, while another sub-system is receiving.

An interesting extension of this concept is to use multiple sub-systems to cover one sector in a cell. This would allow for more than one transmission in each sector per time-slot. Such an arrangement will be denoted a *multi-beam* system. The slot-assigments in the individual sub-systems within a sector are assumed to be coordinated, in order to avoid intra-sector interference.

3. System Model and Performance Measures

In our performance analysis we will make a series of model assumptions. We will confine ourselves to the analysis of a 4-sector system with 90° sectors. We will assume that the radiation patterns of the antennas used have ideal "pencil beam"-shape. The antennas are characterized by their horizontal lobe width angle ϕ_h and the side lobe suppression A_{sl}. The relative received power $P(\phi)$ at the angle ϕ off the main direction may be expressed as

$$P(\phi) = \begin{cases} 1 & |\phi| \le \phi_h/2 \\ \dfrac{1}{A_{sl}} & |\phi| < \phi_h/2 \ . \end{cases} \tag{1}$$

Further, we will assume that the active peripheral stations are randomly (uniformly) distributed over the cell area. A simple propagation model is used in which the received signal level is assumed to be inversely proportional to the distance to the power a, i.e.

$$P_r = \frac{c}{r^a} \ , \tag{2}$$

where a=2 corresponds to line-of-sight communication between elevated antennas. Higher values for a can be used to model non-line-of-sight paths or multipath environments.

In systems with high user-densities, cell sizes can be considered that small that the received power is always much larger than the termal noise power. Instead, the performance of the system is limited by the interference generated by central and peripheral stations in neighboring cells. To simplify our calculations we will assume that the transmission quality is dependent only of the signal-to-interference ratio Γ which may be defined as

$$\Gamma = \frac{P_r}{\sum\limits_{j=1}^{N} I_j} = \frac{1}{\sum\limits_{j=1}^{N} \dfrac{1}{\Gamma_j}}, \tag{3}$$

where P_r denotes the received wanted signal power, wheras I_j denotes the received power from interference source j. We have further introduced

$$\Gamma_j = \frac{P_r}{I_j} \, .$$

We have made the common assumption that the total interference power can be calculated as the sum of the powers from the individual interferers. This assumption holds if we model the interferers as independent random processes.

As our prime performance measure we will use the *relative capacity* S. This quantity is defined as the expected number of links (slots) that are available for communication in a sector in a given slot interval. To greatly simplify calculations, we will here make the assumption that all transmitters also are slot-synchronous. The consequence of this is that the signal-to-interference ratio will be constant throughout the entire duration of a slot. The probability that some slot is available is simply the probability that the signal-to-interference ratio at the receiver is high enough during that slot. For a multi-beam system using M sub-systems per sector, this quantity may be expressed as

$$S = E[\text{no. slot with } \Gamma \geq \gamma_0] \approx$$

$$\approx M \, \Pr[\Gamma \geq \gamma_0]. \qquad (4)$$

Here, Γ denotes the signal-to-interference ratio at some randomly selected receiver. γ_0 denotes the *protection ratio*, i.e the minimal signal-to-interference ratio required to achieve acceptable transmission quaility. The last approximation in the R.H.S. of (4) is valid if the signal-to-interference ratios are independent between simultaneous transmissions in the same sector in a multi-beam system. This assumption will hold with reasonable accuracy as long as simultaneous transmissions within the sector can be kept from "colliding", i.e. if M is not to large (cf section IV). The maximum relative capacity we could imagine is M, corresponding to M independent sub-systems without interference.

Note that S is the expected number of available slots for some given direction. In order to make use of this capacity a sufficient number of active peripheral should be located in this direction, i.e a large number of uniformly distributed peripheral stations is required. Further, an efficient Slot Allocation Algorithm is vital, for the purpose of indentifying interfered slots and reallocating traffic to available slots.

4. Interference Analysis

Let us now investigate the properties of the interference emanating from neighboring cells in somewhat more detail. Since the sub-systems in all cells are assumed to be frame synchronized, two central stations or two peripheral station cannot interfere with each other. Two interference cases remain: a) a receiving peripheral station receives interfering signals from remote central stations ("down-link" interference) or, b) a central station while receiving information from one of its peripherals also receives interfering signals from peripheral stations in other cells ("up-link" interference). From fig 2 it is obvious that these two situations are geometrically equivalent .

The impact of the interfering signals will greatly depend on the position of peripheral and the orientation of the antennas. Is the interferer pointing its antenna toward the receiver and is at the same time located within the "field of vision" of the receiving antenna the interference level is only determined by the propagation law (eq 2). If one of the antennas is pointed in some other direction, the interfering signal will experience an additional attenuation of at least A_{sl}.

Now, let us determine an expression for the signal-to-interference ratio Γ in some arbitrary slot. We will use the geometry and the notation in fig 2. Let ϕ_{cs} and ϕ_{ps} denote the horizontal lobe width of the central and peripheral antennas respectively. We begin by determining G in peripheral station PS in fig 2. The position of PS expressed in polar representation is (r, ϕ). Without loss of generality,

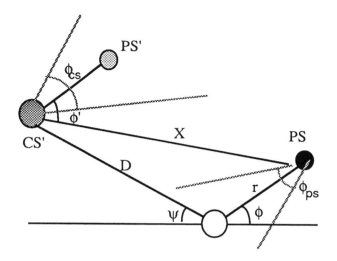

Fig 2. Interference geometry

we will confine ourselves to the case $|\phi|<\pi/6$. We can now find Γ_j, corresponding to interfering central station j. Using the notation in fig 2 we get

$$
\Gamma_j = \begin{cases}
\left(\dfrac{X}{r}\right)^a & r<r_0 \text{ and } |\phi'|<\phi_{cs}/2 \\[3em]
A_{sl}\left(\dfrac{X}{r}\right)^a & r<r_0 \text{ or } |\phi'|<\phi_{cs}/2 \\[3em]
A_{sl}^2\left(\dfrac{X}{r}\right)^a & r>r_0 \text{ and } |\phi'|>\phi_{cs}/2
\end{cases}
\tag{5}
$$

where

$$X^2 = r^2 + D^2 + 2rD\cos(\phi+\psi)$$

$$r_0 = \begin{cases} D\,\dfrac{\sin(\phi+\psi-(\phi_{ps}/2))}{\sin(\phi_{ps}/2)} & |\phi+\psi| \geq \phi_{ps}/2 \\ \\ 0 & |\phi+\psi| < \phi_{ps}/2 \end{cases}$$

The first expression in eq (5) corresponds to the case where both antennas are pointing towards each other. The next corresponds to the case when either the peripheral station is located such that the central station is "visible" but not transmitting in the direction of the peripheral, or the central station is transmitting in the critical direction but it is not "visible" at the peripheral. In the last case both antennas point in some other direction and the interfering signals are attenuated by A_{sl}^2. In the derivation of (5) we assume that $|\phi_{ps}| < \pi/6$, i.e that relatively narrow beam antennas are used.

To compute the total signal-to-interference ratio at a PS we sum the individual signal-to-interference ratios Γ_j over the active interfering central stations (eq 3). From eq (5) it is obvious that the signal-to-interference ratio in the central stations may be computed in exactly the same way. The summation has in this case to be performed over interfering peripheral stations. The statistical properties of the distances are the same, which means that the signal-to-interference ratio in the reverse direction has the same properties as the one computed above.

One important questions remains: in which cases will the interferers point their antennas, such that our peripheral station will be "hit", i.e. when will $|\phi'| < \phi_{cs}/2$? To answer this question, we will assume that the slot allocation in different cells are not coordinated. We further make a pessimistic "full load" assumption, i.e. that all slots are used in the interfering central stations. In the slot of interest, each central station sector sub-system is assumed to communicate with a randomly located peripheral station within its 90°-sector. The angle ϕ'

in fig 2 is thus assumed to be uniformly distributed over the interval[-$\pi/4$, $\pi/4$]. The probability that a peripheral will be hit by the transmission of a single beam system is therefore

$$\Pr[|\phi'| < \phi_{cs}/2] = \frac{2\phi_{cs}}{\pi} \qquad (6)$$

In a multi-beam system, where each 90°-sector is covered by several sub-systems, we may have several transmissions in each sector. These transmissions are, however, assumed to be coordinated, such that no interference will occur in the peripherals. If the stations served in a sector in a given time slot are choosen at random and their number is small we can use the following approximate "hit"-probability for a design with M sub-systems per sector:

$$\Pr[|\phi'| < \phi_{cs}/2] \approx M \frac{2\phi_{cs}}{\pi} \qquad (M << \frac{\pi}{2\phi_{cs}}) \qquad (7)$$

As the number off sub-systems increases, this approximation becomes more and more pessimistic. The maximum number of peripherals that may be served in a sector per time-slot is roughly π/ϕ_{cs}.

5. Numerical Results

In conventional cellular radio systems, the distance to the interfering stations is determined by the frequency repetition pattern. In our study, it turns out that using *the same frequency in all cells* will be perfectly feasible. Evaluating (3), i.e computing all interferer locations and taking the expectation of the expression with respect to the peripheral station locations is rather straigth-forward but extremely teadious. The numerical results in this section are therefore derived by means of numerical and Monte Carlo-methods. Interferers up to a distance of 6 cell radii are considered. We make the rather pessimistic assumption that a) all sub-systems and slots are used (full load), and b) a free-space (line-of-sight) propagation prevails, i.e.

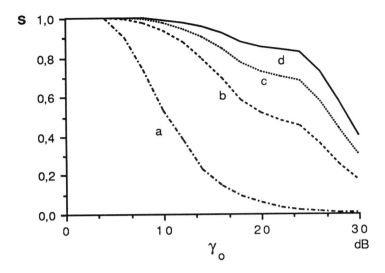

Fig 3. Relative capacity S as function of protection ratio for different central station antenna lobe widths ϕ_{cs} = a) 90° b) 20° c) 10° and d) 5°. ϕ_{ps} = 20° A_{sl} = 20dB

a = 2. The reader is referred to [9] for details of the numerical evaluation.

Figure 3 shows the relative capacity S of a single-beam system as function of the protection ratio γ_0 for various central station antenna lobe widths. The peripheral station antenna is assumed to have fixed a horizontal lobe width of 20°. The "fixed" antenna system indicated in the figure corresponds to a fixed 90° sector antenna. The antennas are assumed to have a sidelobe suppression of 20 dB. We note that for γ_0 around and above the sidelobe suppression the performance is, as expected, dominated by the performance of the peripheral antennas. The graph can be used for the proper design of modulation scheme, i.e. the trade-off between transmission rate and interference sensitivity in order to match the overall throughput and system complexity. Note the dramatic improvement as we go from a fixed central station antenna system to, for example, the 10° antenna. The fixed antenna system, if at all feasible, would basically be confined to

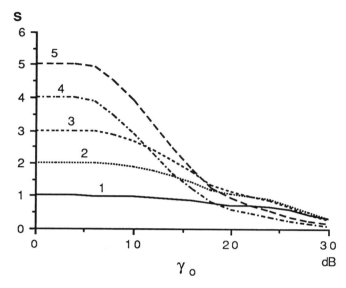

Fig 4. Relative capacity estimate S for a TSR system with multiple sub-system per sector. $\phi_{cs} = 10°$, $\phi_{ps} = 20°$
$A_{sl} = 20dB$

binary modulation schemes or "heavily" coded systems only ($\gamma_o \leq 10$ dB). Otherwise, a more conservative, conventional frequency reuse pattern would have to be used. The latter corresponds of course to a severe loss in capacity. On the other hand, a scanning beam system using a single frequency and a multilevel modulation would be perfectly feasible. As a result, the scanning beam-type system should therefore outperform the fixed antenna system by an order of magnitude.

Figure 4 illustrates the performance of a *multi-beam* system. We see that the relative capacity can be increased considerably compared to the single beam system. As expected, however, the relative improvement decreases with the number of antenna beams. This is due to the fact that it is harder and harder to "squeeze" one more beam into the 90°-sector without causing additional interference. It is interesting to note that for high protection ratios, adding yet another sub-system may in fact lower the overall performance due to the increased interference level. In [9] additional numerical results are found.

6. Discussion

In the paper we have analysed a scanning beam, Time-Space radio link system for metropolitan areas. Rough estimates for the average system capcacity using pessimistic assumption about propagation and interference conditions are found. Even though rather pessimistic assumptions are made, results show that using pencil beams of moderate width dramatically reduces the interference levels in neighboring cells. Contrary to the proposal in [7], a cellular system reusing the same frequency is feasible. This results in large capacity gains compared to existing cellular systems and P-MP radio link schemes. The perfor-mance can be even more enhanced by using a multi-beam scheme, where several transceivers may cover the same area on the same frequency. In addition polarization multiplexing could be used.

An important topic for further research is the design of slot-allocation algorithms. In the study above, the assumption is made that slot assignments are uncoordinated and interference appears "random". This, however, seems to be a pessimistic assumption if some coordination or interference level measurements can be made *before* a slot and an antenna direction is assigned. This problem should be closely related to the interference based dynamic channel allocation problem in cellular telephone systems.

Other cellular structures, more suited for directional antenna systems should constitute another interesting problem complex. Using hexagonal cell structures in combination with 90°-sector systems may not be the optimum choice. Further, the disadvantage peripheral stations experience when located on one of the symmetry-lines of the cell pattern should generate some interest to study non-symmetric or maybe even randomized cell-patterns.

7. References

[1] Le-Ngoc, T., Zavitz, H.J., "A point to multipoint digital microvawe link for rural applications", Com. Eng. Int, April 1985.

[2] Hall, H., Westerlund, L., Åhl, K-A, "Rural telephony with radio concentrators", TELE, Swedish Telecom, No 2, 1983.

[3] Evagora, A., Finnie, G., "The Future of the Local Loop", Communications Week International, Oct 11, 1991.

[4] Swales, S.C., Beach, M., Edwards, D.J., "Multi-beam Adaptive Antennas for Cellular Land Mobile Radio", *Proc. Veh Tech Conf., VTC-89,* San Francisco, 1989.

[5] Reudink, D.O., Yeh,Y.S., "A scanning spot beam satellite system", *Bell Syst. Tech. J.,* vol 56, no. 8, Oct 1977.

[6] Acampora, A.S, Dragone, C., Reudink, D.O, "A satellite system with limited scan spot beams", *IEEE Trans. Commun.,* vol COM-27,Oct 1979.

[7] Acampora, A.S, Chu, T-S., Dragone, C., Gans, M.J., "A Metropolitan Area Radio System Using Scanning Pencil Beams", *IEEE Trans. Commun.,* vol COM-39, no.1, jan 1991.

[8] Åhl, K-A., Nelson,J., Lindfors, K., "Method and device in digital communication network", Int. pat. appl., WO91/06162, May 1991.

[9] Zander, J., Åhl, K-A., "On the capacity of Time-Space Switched Cellular Radio Link Systems for Metropolitan Area Networks", *Technical Report*, TRITA-TTT-9117, Royal Inst. Technology, ISSN-0280-4492, Oct 1991.

Network Contention Issues in the Design of Very Large Meteor Scatter Networks for Vehicle Tracking and Communication

J.A. Weitzen[*], J.D. Larsen, and R.S. Mawrey

Meteor Communication Corporation
6020 South 190'th Street
Kent Washington 98032

Abstract

Meteor scatter communication operating in the low VHF band, using the ionized trails of meteors entering the atmosphere as "free" naturally occurring satellite channels, represents a low cost alternative technology for automatic vehicle location (AVL) and vehicle communication. One of the key issues in the design of a large system is contention within the network due to the presence of more than one vehicle within the footprint created by each meteor trail. This paper describes a technique for predicting the probability of contention in a dense meteor scatter network. It is shown that with proper protocol design, networks with 4000 or more vehicles per frequency allocation can be supported by a single base station.

I Introduction

A practical, affordable system for communicating with mobile platforms such as cars, trucks, railroad cars, and marine vessels at a distance has many uses. It permits a trucking company to know the current locations of all its vehicles and to reroute individual vehicles or otherwise optimize the use of its fleet [10]. A related use is the tracking of hazardous material carriers, such as nuclear waste or ammunition trucks, to improve system safety. Other uses include vehicle probing to provide adaptive traffic management.

Past and current attempts to solve this problem used conventional communication techniques such as HF, VHF/UHF line-of-sight, or satellites. These techniques all have their individual disadvantages and advantages. HF is unreliable, subject to fading, and requires multiple frequency allocations and adaptive techniques to manage frequency changes. It also tends to require sequential polling of the individual vehicles. For these reasons its applicability to very large networks is limited.

---

[*] J.A. Weitzen is also Associate Professor of Electrical Engineering at University of Massachusetts Lowell

Conventional UHF and VHF are essentially line-of-sight, and hence coverage of a large area requires many base stations and complicated hand-off and contention protocols. This is a practical method as shown by current developments in cellular radio technology. Cellular based system do not provide full national coverage because in remote areas and away from the interstate highways, there is inadequate subscriber density for cost effectiveness.

Satellites are currently the favored method, and several companies are fielding systems using different satellite technologies [4]. Satellites have the primary advantage of inherent wide area coverage, but have the disadvantage of the high cost of launching the satellite and require sophisticated mobile terminals. They also are subject to single-point failure of the entire system unless redundant satellites are used, again adding to the cost.

Meteor burst communication technology, using the ionized trails of meteors entering the atmosphere as free, naturally occurring satellites has not yet been effectively applied to the problem of vehicle communication. Current operational meteor burst systems are, by and large, either point-to-point systems using high gain antennas, providing communication in the 10 to 500 words per minute range, or are wide area systems with low gain antennas, which serve widely dispersed remote units with wait times on the order of minutes to an hour [6].

Meteor scatter has a number of advantages as a low cost alternative to cellular or satellite based systems. It is an extended range beyond line of sight propagation mechanism so that fewer ground stations are needed to service a wide area. The entire continental U.S. could covered by a network consisting of less than 100 ground stations. This provides for a lower cost infrastructure investment than conventional systems. The mobile terminal equipment is relatively inexpensive which further lowers the cost of the system and the free meteors eliminate the need for an expensive satellite transponder. Because it does not depend on the unreliable and dynamic ionosphere for propagation, a single frequency can be used. Finally, the inherent space division multiple access (SDMA) or spatial multiplexing provided by the relatively small ground illumination footprint allows frequency reuse so that a small allocation (20-40 kHz in the 40-50 MHz VHF band) can serve a very large number of users.

Because of the wide area coverage and low cost, meteor scatter represents an attractive low cost alternative to satellite based systems for communication of short packetized messages to a very large number of mobile users. The paper by Larsen et al, [6] discusses many of the general design requirements for a commercial or military communication system using meteor scatter communication for vehicle tracking. This paper concentrates on the issue of contention within the network which is one of the key issues in the design of a commercial meteor scatter communication network.

II The Meteor Scatter Footprint Effect

One of the unique advantages of meteor scatter propagation is the so called "footprint" effect. Because meteor communications uses as the propagation mechanism scatter from small tubular clouds of electrons, the region on the ground which is illuminated by a given trail is much smaller than that for conventional communication mechanisms. In the past this effect has been viewed as providing a degree of inherent privacy or resistance to interception [3, 8, 9, 11, 13]. From the view of large network design, the footprint effect provides an inherent degree of spatial multiplexing or space division multiple access (SDMA) which acts to prevent contention between users as is a problem in conventional ALOHA packet radio networks. This paper describes a technique to predict the number of remote units in a network which are connected at any one time. The results of this analysis have been used in the design of the protocols which control the Meteor Communications Corporation (MCC) trucking network currently under development.

The small ground illumination region occurs as the result of two fundamental properties of the meteor scatter propagation mechanism. The first condition is the classic tangency theorem which requires that the line forming the trail be tangent to an ellipsoid of revolution with foci at the receiver and the transmitter. It has been shown that for a given line and a receiver and transmitter, a point of tangency exists along the line. For a set of nodes with a common transmitter, the points of tangency will lie at different points along the line forming the trail. The second requirement is that there be significant ionization at the point of tangency.

Figure 1 illustrates how the combination of the two conditions gives rise to a dynamic ground illumination region. The confusion between the instantaneous and trail footprint exists because the velocity of the meteor forming the trail is finite. The region of ionization begins at the head of the moving meteor. Immediately after formation, the trail begins to expand due to diffusion and the received signal decays. Beyond some distance behind the meteor head, the electron density has decayed to a point at which communication is no longer supported by the trail. Nodes for which the point of tangency lie within this region are said to be in the instantaneous footprint of the trail.

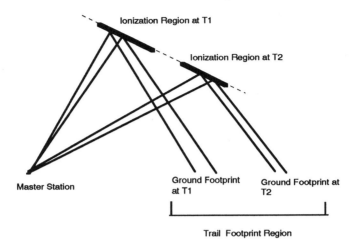

Figure 1. Motion of the ground illumination region due to trail formation and decay.

As the meteor continues along its path, new nodes become connected as the region of significant ionization coincides with their points of tangency. Other nodes become disconnected when the region of ionization moves beyond their points of tangency. If the meteor velocity were infinite or semi-infinite, there would be no difference between the trail footprint and the instantaneous footprint. The instantaneous footprint is a subset of the trail footprint.

Calculation of Individual Trail Footprints

To calculate the region on the ground which is illuminated by individual trails, we use a modified version of the technique outlined in [13]. Assume the meteors forming trails have an average heliocentric velocity of 35 km/sec [2, 7]. The region of ionization exists for a distance equivalent to two underdense time constants from the head of the meteor. For an underdense trail, this is equivalent to 10 dB less signal than at the head.

Consider the sky and secondary or ground coordinate systems shown in figure 2. The sky system is a three dimensional Cartesian system with origin at the midpoint of the chord connecting the primary transmitter and primary receiver which are assumed to lie on a spherical earth. The secondary or ground coordinate system is a two dimensional Cartesian system with origin at the primary receiver in the sky system.

Consider a meteor trail with mid-point at (x,y,z) and with direction vector $\vec{T}$ which is observed at the primary receiver at $(0,0)$ in the ground coordinate system.

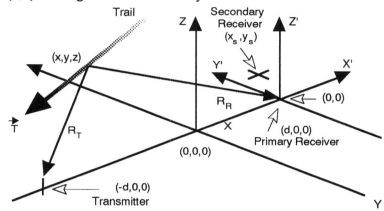

Figure 2 Primary (sky) and secondary (ground) coordinate systems for footprint analysis.

Moving on the ground away from the primary receiver, the point on the trail at which the tangency conditions are satisfied for the new receiver location moves. If the new tangent point moves far enough from the original receiver so that the new point is no longer in the ionized region of the trail, the new receiver location will not be illuminated by the trail.

Let (x_t,y_t,z_t) be the point of tangency for a secondary receiver at (x_s,y_s) assuming a trail with mid-point (x,y,z), length L, and direction vector $\vec{T}$. The mid-point (x,y,z) represents the tangent point when the receiver is located at $(0,0)$ in the ground coordinate system so that the trail $\vec{T}$ creates a channel from the transmitter to the primary receiver.

The tangency condition for specular scatter at any ground location can be expressed as

$$\vec{N}\cdot\vec{T}=0 \tag{1}$$

where $\vec{N}$ is the normal vector to the ellipsoid at the point (x_t, y_t, z_t). The requirement that the new tangent point lie within the region of ionization is expressed as

$$|(x,y,z)-(x_t,y_t,z_t)| \leq \frac{L}{2} \tag{2}$$

Express the trail direction vector in Cartesian coordinates as

$$\vec{T}=a\hat{a}_x+b\hat{a}_y+c\,\hat{a}_z \tag{3}$$

A point on the trail (x_t, y_t, z_t) satisfies the parametric equations for a straight line with direction vector $\vec{T}$ given by

$$\frac{x-x_t}{a} = \frac{y-y_t}{b} = \frac{z-z_t}{c} \tag{4}$$

Let

$$\vec{R}_T=(x_t+d)\hat{a}_x+y_t\,\hat{a}_y+z_t\hat{a}_z \tag{5}$$

and

$$\vec{R}_R=(x_t-d-x_s)\,\hat{a}_x+(y_t-y_s)\,\hat{a}_y+z_t\,\hat{a}_z$$

represent vectors from the tangent point (x_t, y_t, z_t) to the transmitter $(-d,0,0)$ and a secondary receiver at $(d+x_s, y_s, z_s)$ respectively. Rudie showed that the normal to an ellipsoid with foci at receiver and transmitter is given by

$$\vec{N}=0.5\left[\frac{\vec{R}_T}{|\vec{R}_T|}+\frac{\vec{R}_R}{|\vec{R}_R|}\right] \tag{6}$$

Substituting (3),(5), and (6) into (1) yields

$$|\vec{R}_T|^2(a(x_t-d-x_s)+b(y_t-y_s)+cz_t)=-|\vec{R}_R|^2(a(x_t+d)+b(y_t)+cz_t) \tag{7}$$

Equations (4) and (7) are combined to solve for the tangent point (x_t, y_t, z_t) using numerical methods. If (2) is satisfied, then a secondary receiver at (x_s, y_s) can receive a signal received at $(0,0)$.

Let $C_f(x_s, y_s, x, y, z, \alpha, \beta)$ represent the ground illumination region of a trail with mid-point at (x,y,z) and orientation (α, β) as a function of location on the ground relative to the primary receiver. For mathematical expediency, define $C_f(x_s,y_s,x,y,z,\alpha,\beta)$ as a binary function: "1" if the point (x_s,y_s) is illuminated and "0" if not illuminated by the given trail. The size and shape of the ground illumination footprint is a function of the orientation of the trail and its distance from the receiver. The size and shape of both the "instantaneous" and "trail" footprints are influenced by the geometry of the link and the orientation of the trail.

To illustrate how the trail footprint is a function of orientation of the trail, assume a 660 km link. Consider first trails located at the mid-point of the link oriented along and normal to the great circle path from transmitter to receiver. See [13] for more information on the size and shape of meteor trail footprints.

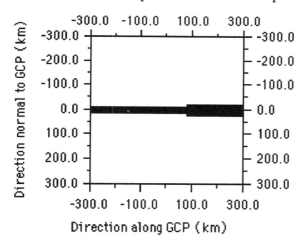

Figure 3. Trail footprint due to trail oriented normal to great circle plane at the link midpoint [13].

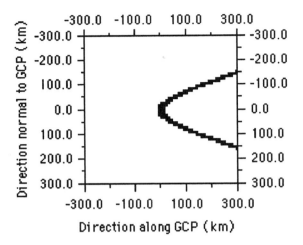

Figure 4. Trail footprint due to trail oriented along great circle plane at the link midpoint [13].

The size of the footprint is also greatly influenced by its location in the common volume relative to the link geometry. Assuming the same link consider trails which are located approximately above the transmitter (figure 5) and above the receiver (figure 6). The trail located above the transmitter will have a much larger footprint and less spatial multiplexing than a trail located above the receiver. This phenomena provides additional incentive for the use of end-path illumination proposed by [5].

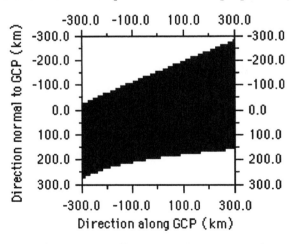

Figure 5. Footprint due to trail located above transmitter [13]

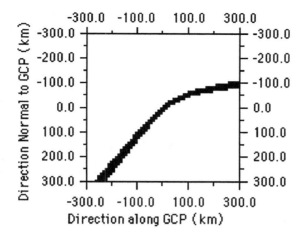

Figure 6 Footprint due to meteor trail located above receiver [13].

III Computing Probability Of k nodes Receiving a Packet

To compute the probability of k nodes receiving a given message transmitted over the same trail p(k), the following set of calculations are performed. First, we iterate over all points in the meteor region of sky coordinate system at a fixed z corresponding to 95 km altitude). Next, at each point in the sky, we iterate over a number meteor trail orientations in azimuth and elevation (α, β). Trails of 40 km total length, centered at the point (x,y,z) in the primary coordinate system are considered in the analysis. For each node in the network, we compute the point of tangency on the trail using the technique described previously.

The model assumes a meteor with heliocentric velocity of 35 km/sec which begins forming the trail at t=0. The time the meteor forming the trail requires to traverse the 40 km trail is divided into 40 ms intervals (message packet length) and the time at which each node is first illuminated by the trail is computed.

Iterating over all the 40 ms intervals we count how many nodes are illuminated at this time or within 2 underdense time constants of this time. Two underdense time constants corresponds to approximately 10 dB signal decay assuming the underdense model. This is done for each 40 ms interval and the distribution of p(k) is computed for the individual trail orientation.

Next, the p(k) distribution for the given trail orientation is weighted by the relative occurrence of trails at this orientation and location in the common volume using the METPRED model [12]. After looping over all locations within the common volume we normalize to determine the relative probability of k responses.

Predictions of the model were compared to data from a USAF experiment with up to 33 nodes. The results were favorable and analysis of the data and some preliminary results are described in [1].

To illustrate the use of the technique described in this paper in the design of the network protocols for a very large meteor scatter network, consider a scenario with 2000 trucks per base station located within a ± 30 degree arc which is the coverage area of a typical base station. The coverage range for meteor scatter propagation extends from 200 km to 1100 km from the base station. At shorter ranges, connectivity is provided by line-of-sight or diffraction propagation and the system resembles a typical slotted ALOHA system. The distance from the master station to the center of the grid shown is 650 km. Vehicle locations are chosen at random as shown in figure 7. The average vehicle density is approximately 1 vehicle per 130 km^2 which is typical of that encountered in a real application. In addition there are cases such as truck stops in which there might be many vehicles in close proximity.

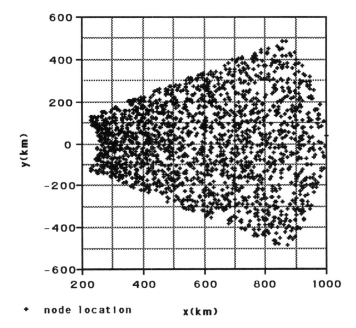

Figure 7. Deployment of 2000 vehicles within typical coverage area of a base station

In the following two figures we plot probability of k receivers residing within the instantaneous footprint, p(k), as a function of k for a control (no additional protocol control) and 5 different protocols which we label 1 through 5 . For each protocol, the probability of more than 1 unit responding (cprob) and the average number of units responding (kavg) are calculated.

Importantly, note that the predictions for very large networks represent upper bounds since all units are assumed identical. As the network density grows, effects due to slight differences between units such as antenna patters and local noise conditions can further reduce the probability of multiple responses.

It is observed in the figures that the spatial multiplexing in the meteor channel alone is inadequate to prevent contention for a very dense network, however, with additional controls on response built into the network protocols, very large network densities can

154

be supported. Design of networks capable of supporting in excess of 4000 nodes per base station is currently under way.

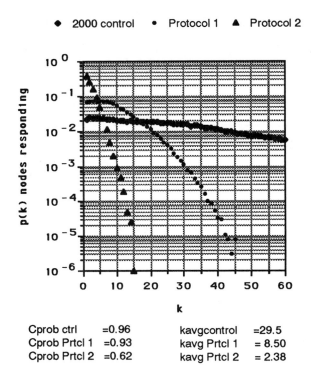

Figure 8. Upper bound on P(k) for 2000 nodes shown in figure 7. Control and protocols 1 and 2.

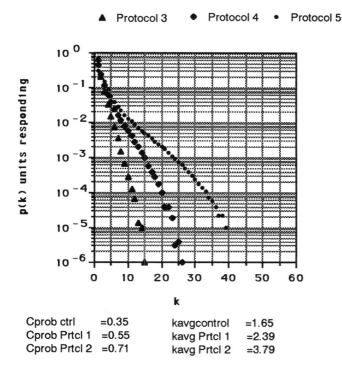

Figure 9. Upper bound on p(k) for 2000 nodes shown in figure 7. Protocols 3,4 and 5.

IV Conclusions

A technique has been developed for predicting the probability of more than one node (vehicle) in the footprint of an individual trail. Preliminary analysis shows the model to be accurate when compared to data from a large government sponsored experiment. Extension to the problem of dense networks with up to 4000 nodes per master station shows that careful design of the protocols coupled with the natural spatial multiplexing in the channel can reduce the probability of contention to acceptable levels. While there appears to be inadequate inherent spatial multiplexing in the channel to support the densities required, with careful design of the network control protocols, meteor scatter represents a viable low cost alternative to satellites for the emerging market of vehicle tracking and positioning.

156

V. References

1. Bennet, T. D., L. L. Baggerly and J. E. MacCarthy, "Some measures of meteor burst footprint", *Proc MILCOM*, Washington, D.C., Nov 4-7, 1991

2. Davies, J. G., "Radio observation of meteors," *Advances in Electronics and Electron Physics*, Marton ed. 1957 , pp 95-127, Academic Press, New York.

3. Ince, A. N., "Interception of signals transmitted via meteor trails", *Proc AGARD Symposium "Aspects of Electromagnetic Scattering in Radio Communications"*, 3-7 October, Cambridge Ma, 1977

4. Jenks, A., Qualcomm, A big winner after Geostar's Crash. Washington Technology. 8, 1991.

5. Larsen, J. D. and R. S. Mawrey, "Feasibility of End-path Illumination in Meteor Scatter", *Proc IEEE MILCOM*, Washington, D.C. Nov 4-7, 1991

6. Larsen, J. D., R. S. Mawrey and F. L. Anderson, "Military Vehicle Communication via Meteor Burst", *Proc IEEE MILCOM*, Washington, D.C., Nov 4-7, 1991

7. McKinley, D. W. R., "Meteor Science and Engineering." 1961 McGraw-Hill Book Company, New York, NY.

8. Mui, S. Y., "Meteor Trail Footprint Statistics", *Proc IEEE Military Communications Conference*, Monterey, CA, 1990

9. Niedenfuhr, F. E., "An Analytical model of meteor burst communication reception zones", *Proc Proceedings of the SHAPE meteor burst communications symposium*, The Hague, Netherlands, November 4-5, 1987

10. Stadden, K. and P. McCullough, To Locate and Communicate. Heavy Duty Trucking. 55-70, 1987.

11. Vogan, E. L. and P. A. Forsythe, "Privacy in System "JANET"", Project Report, Defence Research Telecommunications Establishment, 12-3-4, 1953

12. Weitzen, J. A., "Predicting the Arrival of Meteors Useful for Meteor Burst Communication", *Radio Science*, vol. 21, No 6, November-December, pp 1009-1020, 1986.

13. Weitzen, J. A., "A Study of Ground Illumination Pattern of Meteor Scatter Communication", *IEEE Trans on Comm*, vol. Com-38, No 4., Apr., pp pp 426-431, 1990.

TELETRAFFIC MODELS FOR URBAN AND SUBURBAN MICROCELLS: CELL SIZES AND HANDOFF RATES

Sanjiv Nanda
Room 3M-317
AT&T Bell Laboratories
Holmdel NJ 07733-3030

(908) 949-1759
nanda@hocus.att.com

Abstract

As cellular systems evolve into personal communication networks, teletraffic modeling of users will become crucial. Teletraffic models are required for cellular system layout and planning and to evaluate tradeoffs in system design issues. In this paper, we begin with a study of handoff rates in evolving cellular systems. Using simple geometry, we show that in macrocells with homogeneous traffic the handoff rate per call increases only as the square-root of the increase in the call density. The situation is different in microcells where individual traffic paths become important and the homogeneous traffic model does not apply. Under smooth flow, the handoff rate along a traffic path goes up linearly with the number of new boundaries intersecting the path. Hence, in urban and highway microcells the handoff rate per call increases linearly with increasing call

158

density. We propose a parametric model based on results from vehicular traffic theory, that applies to vehicles and pedestrians in urban, suburban and highway cellular systems. The model helps quantify various design tradeoffs. We conclude that microcells are unnecessary in urban environments if the pedestrian penetration is small. An architecture with microcells and overlaid macrocells will be necessary only when a majority of teletraffic is pedestrian generated.

1. Introduction

Teletraffic theory has concerned itself with the study of blocking of random call arrivals at a switch. In cellular systems, the number of telephones in a base station or switch coverage area is itself random and time varying. As users with mobile telephones move between cells, large variations in the number of telephones in a base station coverage area occur in a matter of minutes. Therefore, teletraffic modeling for cellular systems must incorporate mobility. Variations in vehicular traffic, build-up of traffic jams and movement of pedestrians, all determine the number of mobile telephones in a cell, their speed and their direction of movement. These relate directly to the call arrival rate and handoff rate. The mobility model would also help estimate other mobility management messaging traffic that is required, including location updates, autonomous registration and paging. To be useful for modeling the teletraffic and control messages in a cellular system, the mobility model should incorporate (i) the size and shape of cells and (ii) the speed and direction of mobiles. Our goal is to construct a model with a small number of critical parameters.

In this paper, we consider the crossing of area boundaries as one aspect of mobility modeling. This has relevance to the handoff rates, rates of location updates and other mobility management issues. We begin with a study of handoff rates in evolving cellular systems. Although, in most contexts *rate* refers to a time average, for the purpose of this paper we will use the term handoff rate to refer to the mean number of *handoffs per call*.

In Sections 2 and 3 we use geometric models to predict handoff rates per call as cell shapes and sizes are varied. A similar approach is used in [2] explicitly, and implicitly by various other authors. We find that the handoff rate per call increases as the *square root* of the increase in the number of cells per unit area. This result holds under the following assumptions: (i) the user mobility does not change as the traffic increases and the cell size becomes smaller, and (ii) the cell sizes are large compared to the individual traffic paths (highways and streets), so that mobile travel may be considered to be distributed randomly in the plane. In Section 3, we find that if the direction of flow of mobiles is not uniformly distributed over $[0, 2\pi)$, then it is desirable to shape cells so that the longer boundaries are parallel to the larger flows.

In Section 4 we prove two fundamental results: (i) By appealing to ergodic theory, we show that the mean handoff rate is just the mean call duration divided by the mobile's mean sojourn time in a cell, and (ii) The increase in handoff rate, as old cells are split and new ones added, depends only on the number of traffic paths intersected by the new boundaries introduced. As cell sizes get smaller, so that a small number of individual traffic paths dominate a cell, we can no longer assume that mobile travel is uniformly distributed in a cell. If the user mobility remains the same and each cell contains exactly one traffic path (e.g. highway microcells), we find that the handoff rate increases *linearly* with the increase in the number of cells per unit area.

In Section 5, using vehicular traffic theory (see for example [3]), we devise a parametric model that will be applicable to personal communication networks of the future. Vehicular traffic theory was applied to construct a teletraffic model for highway microcells in [4]. Although, developing a teletraffic model was not the primary goal of that paper, trade-offs between microcell and overlaid macrocell size, when traffic is free-flowing or congested, were discussed there. In this paper we consider the more general scenario of modeling teletraffic in an urban or suburban area, as parametrized by the roadway density in lane-kilometers per square kilometer. Thus, highways, arterial roads and urban streets are included in our model. In addition to free-flowing and congested vehicles, our model includes pedestrians. In Section 5, we derive results for mean length of mobile (pedestrian or vehicle) travel in a macrocell and in a microcell.

In Section 6, using realistic values of the parameters we find that high user densities that require very small cells will be reached only with high levels of pedestrian penetration. Our assumption in Sections 2-4 above, of unchanged user mobility at high user density, is invalid for PCN's. In fact, high user densities will be characterized by a decrease in the average mobility. The above results for square-root or linear increase in handoff rate still apply, but only to the users with mobility similar to today's cellular users. But, with increasing user density, the mean handoff rate will be dominated by the pedestrian users with low mobility. Hence, even as the high mobility users will experience large increases in handoff rate, the mean handoff rate will increase slowly. It may even decrease, depending on pedestrian mobility and pedestrian penetration.

Our conclusions are in Section 7, where we also discuss the relationship of our approach to the vehicular traffic simulator developed at WINLAB [5].

2. Geometric Modeling of Cells - Size

As today's cellular system evolves into smaller cells, simple geometry can be used to predict handoff rates in the smaller cells. Since coverage areas are often designed to carry equal amounts of teletraffic, equal-sized cells, as shown in Figure 1, carrying homogeneous traffic is a useful abstraction.

We make the following assumptions:
1. Mobiles are distributed uniformly in the region.
2. Direction of travel of mobiles is uniformly distributed over $[0, 2\pi)$.

Under these assumptions, the handoff rate is directly proportional to the total boundary length. We use this result to calculate the increase in handoff rate as the cell size decreases.

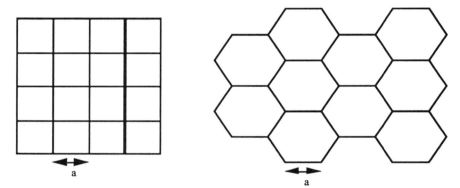

Figure 1 Square and Hexagonal cellular grids

Let us assume that the initial system has N channels per cell, with each cell a square of side a. The current demand is adequately served with a given Grade of Service (GOS) by the channel density, ρ (channels per unit area) $= N/a^2$. As the demand increases, we require a higher channel density ρ_{new} (channels per unit area) for the same GOS, and additional cells and channels must be provisioned. By considering channels per unit area instead of channels per cell, we are imprecise about trunking efficiency. New technology and spectrum allocations increase the channels per cell by a factor G, to GN. With the existing cells this gives a channel density of $\rho_1 = GN/a^2$, which is still short of the required ρ_{new}. The cell size must be reduced from squares of side a, to squares of side a/F, so that $\rho_{new} = GN/(a/F)^2$, and the number of channels per unit area increases by a factor GF^2. The corresponding expressions for the hexagonal lattice, with hexagons of side a, are $N/(1.5\sqrt{3}\ a^2)$ and $GF^2N/(1.5\sqrt{3}\ a^2)$.

The handoff rate also increases as the number of calls per unit area goes up. Again, considering the square lattice, the total cell boundary length per unit area increases from $4a/a^2$ to $4aF/a^2$, that is, by a factor F. With the increase in call density discussed above, the increase in handoff density is by a factor of GF^3! The number of calls per unit area increases as GF^2 and the number of handoffs per unit area increases as GF^3, therefore, the handoff rate per call increases as F.

The development here assumes that there is no change in user mobility in the increased demand, that is, the behavior of new mobiles is no different from existing ones. If the existing demand is dominated by vehicular mobiles, and the increased demand comes from pedestrian hand-helds, then this assumption does not apply. This aspect is discussed later in the paper.

3. Geometric Modeling of Cells - Shape

We next consider the effect of cell shape on the handoff rate. For homogeneous traffic, the handoff rate depends only on the length of the boundary per unit area. As seen above, the boundary length per unit area for the hexagonal lattice is a factor of $\sqrt{3}$ smaller than for the square lattice. The hexagonal lattice is the best by this measure. Although the circle has a boundary length per unit area of $2/a$ (a factor of 2 smaller than the square lattice, circles do not tessellate).

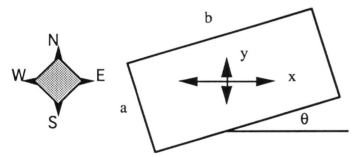

Figure 2 Rectangular cell oriented at an angle from the axes

Next, consider a situation when the traffic is directional. Then, the handoff rate depends, not only on the length of the boundary, but its orientation as well. Suppose that the number of vehicles crossing a unit length of boundary, oriented N-S, is x; and the number of vehicles crossing an E-W oriented boundary is y. And consider rectangular cells of area A. Then if we choose a rectangle which is oriented at an angle θ from the E-W direction (Figure 2), with sides a, b (so that ab = A), the handoff rate H, for the cell, is given by

$$H = a(x \cos \theta + y \sin \theta) + b(x \sin \theta + y \cos \theta) \qquad (1)$$

With the cell area fixed at A, the number of calls in the cell is fixed; then for fixed θ, to minimize the handoff rate H, we must choose,

$$a = \sqrt{A}.\sqrt{\frac{x \sin\theta + y \cos\theta}{x \cos\theta + y \sin\theta}}, \qquad b = \sqrt{A}.\sqrt{\frac{x \cos\theta + y \sin\theta}{x \sin\theta + y \cos\theta}}$$

which gives,

$$H = 2\sqrt{A}\sqrt{xy + (x^2+y^2)\sin\theta\cos\theta}$$

This is minimum for $\theta = 0$, so that the handoff rate is minimized if the rectangle is oriented in the E-W direction, with the boundaries normal to the flows. With $\theta = 0$,

$$a^2 = Ay/x, \qquad b^2 = Ax/y$$

$$\text{and} \quad H = 2\sqrt{Axy} \qquad (2)$$

Notice that this gives $(b/a) = (x/y)$, which is intuitive, since it suggests that we must put the longer boundaries in the direction that has a smaller boundary crossing rate. In particular, this argument leads to linear highway microcells.

4. Fundamental Results on Mean Handoff Rates

4.1 Handoff Rate from Call Holding Time and Cell Sojourn Time

Assume an exponentially distributed call holding time of τ seconds, with mean $1/\mu$. As a mobile traverses cells assume that its sojourn times t_1, t_2, ..., are independent, identically distributed random variables with the same mean $1/\gamma$. We first show that in this special case, the average handoff rate is γ/μ. We have shown that this result holds for arbitrary call holding time distributions and stationary cell sojourn time distributions.

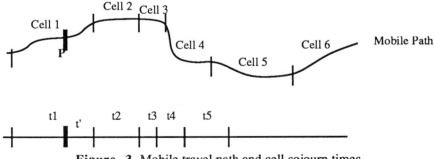

Figure 3 Mobile travel path and cell sojourn times

Consider a mobile traversing a path shown in Figure 3, with consecutive cell sojourn times t_1, t_2, ... Assume that a call begins at P so that the remaining sojourn time in the cell is t'. Let H be the number of handoffs per call, and we are interested in E[H]. Given the mobile's sojourn times in the cells, the number of handoffs experienced by the call depends on the call holding time, which is exponentially distributed with mean $1/\mu$. Hence,

$$
\begin{aligned}
E[H \mid t', t_2, t_3, \ldots] &= 1. \int_{t'}^{t'+t_2} \mu e^{-\mu x} dx + 2. \int_{t'+t_2}^{t'+t_2+t_3} \mu e^{-\mu x} dx + 3. \int_{t'+t_2+t_3}^{t'+t_2+t_3+t_4} \mu e^{-\mu x} dx + \ldots \\
&= \int_{t'}^{\infty} \mu e^{-\mu x} dx + \int_{t'+t_2}^{\infty} \mu e^{-\mu x} dx + \int_{t'+t_2+t_3}^{\infty} \mu e^{-\mu x} dx + \ldots \\
&= e^{-\mu t'} [1 + e^{-\mu t_2} [1 + e^{-\mu t_3} [1 + \ldots
\end{aligned}
\tag{3}
$$

Assume that the t_i are identically distributed with probability density function $g(t)$, and the Laplace Transform of $g(t)$ is given by $G^*(s)$, then

$$
\begin{aligned}
E[H \mid t'] &= e^{-\mu t'} [1 + G^*(\mu)[1 + G^*(\mu)[1 + \ldots \\
&= \frac{e^{-\mu t'}}{1 - G^*(\mu)}
\end{aligned}
\tag{4}
$$

To obtain E[H], we need the distribution of t' the remaining sojourn time in cell 1. The problem is simplified if the t_i are exponentially distributed, then t' has the same distribution as t_i due to the memoryless property of the exponential distribution. In general, let t' have probability density function $r(t)$, with Laplace transform $R^*(s)$. Then taking the expectation of (4),

$$
E[H] = R^*(\mu) / (1 - G^*(\mu))
\tag{5}
$$

Using the results for residual service time, see for example [6], we have that

$$
R^*(s) = \frac{1 - G^*(s)}{s/\gamma}
\tag{6}
$$

And therefore, substituting (6) in (5),

$$
E[H] = \gamma/\mu.
\tag{7}
$$

Hence, the mean handoff rate is given by the ratio of the mean call holding time to the mean cell sojourn time, which is what one would intuitively expect.

4.2 Handoff Rate Increase due to Cell Splitting
In this section, we begin to relax the assumptions of homogeneity used in the previous sections. In Figure 4 we show a section of a traffic path (e.g. a road, a

164

highway or a city street) that crosses a cell, intersecting the boundary at two points C and D. If the mean flow rate on this section of the traffic path is constant (q vehicles per second), then the number of handoffs per second is just the mean flow rate of mobiles that are in-call. We assume that the mean number of vehicles that have calls in progress at any instant is α, which we define as the in-call probability.

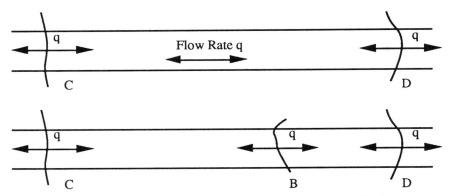

Figure 4 Introducing new boundaries intersecting a traffic flow

Hence, there are αq handoffs per second at each intersection of the cell boundary with the traffic path. If cell splitting introduces a new boundary at any point B along this section of traffic path, there will be αq handoffs per second at the new intersection also. Hence, the introduction of the new boundary doubles the number of handoffs per second for this section of the traffic path. If we consider individual mobiles we get the following result for handoffs per call. For fixed mobility, the handoff rate increases linearly with the number of new cell boundaries that intersect any traffic path.

The results obtained in Sections 2 and 3, assuming the homogeneity of vehicular traffic apply to situations where each cell consists of a large number of traffic paths with random orientation. Cell splitting produces smaller cells that have similar characteristics, and the geometric results apply. As cellular systems evolve to smaller cells, the homogeneous traffic assumption breaks down; the distinct paths in each cell, their orientation and traffic flow, become important. Projection of handoff rates to these small cells must be done using the results obtained in this section, by counting the intersections of traffic paths by new cell boundaries.

Example: Consider the map consisting of major arteries in a geographical area, shown in Figure 5. To simplify the argument assume that cellular traffic and the handoffs are dominated by the traffic on the major arteries and that the flow along each major artery is the same. In the original cell the major arteries intersect the cell boundary at 12 points.

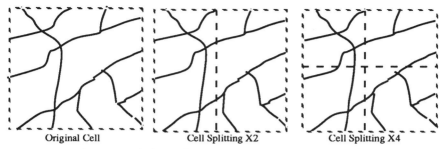

Original Cell Cell Splitting X2 Cell Splitting X4

Figure 5 Schematic map to demonstrate cell splitting

Consider the first cell splitting by a factor of 2. The handoff rate would increase by a factor of 2 on the paths along which a new boundary is created while staying the same on the paths that are not intersected by the new boundary. In the example shown the left split cell has 8 points where the major arteries intersect a boundary while the right split cell also has 8 such points. Thus, the handoff rate increases by a factor of 16/12, which is 1.33. We point out that a similar split with a horizontal boundary would create 19 such boundary crossing points corresponding to a handoff rate increase of 1.58. (This also demonstrates the need for careful planning required for cell splitting). Next, consider the cell split into 4. We now find that the four split cells each have either 5 or 6 points where the arteries cross boundaries, with a total of 22 such crossings and a handoff rate increase by a factor of 1.83.

The main result of this section is that the handoff rate increases linearly as new boundaries are introduced intersecting traffic paths. Our discussion here suggests that this result is consistent with the result of Section 2, that the increase in handoff rate is only as the square-root of the increase in cells per unit area. The square-root result holds when cells consist of a large number of randomly oriented traffic paths. In microcells, dominated by a few or just one traffic path, the increase in handoff rate is linear with the increase in cells per unit area.

5. A Parametric Model

As user densities increase, the current vehicular to pedestrian traffic mix will change and as discussed in the last section, the cell geometry will changes as well. Hence, homogeneous traffic models will be inapplicable. Handoff rates for future personal communication networks cannot be projected directly from current measurements and geometric models.

166

We propose an abstract teletraffic model suitable for PCN's, that describes call arrivals and handoffs from vehicles and pedestrians in urban, suburban and highway settings. We identify key parameters that lead to general models, keeping in mind that the chosen parameters should have physical significance to system designers. Before we describe the model, we outline some basic results from vehicular traffic theory.

5.1 Results from Vehicular Traffic Theory [3]

Vehicular Traffic Theory has developed over the last thirty years and has been applied for studies of highway and street capacities and to design roadway systems that offer congestion-free flow of traffic. Reference [3] is an excellent tutorial on the subject. Some of the empirical flow versus density relationships are important for our mobility modeling, and we discuss them here.

The Fundamental Equation of vehicular traffic flows relates the speed of vehicles on a traffic path, s (in mph), to the density, k (vehicles per mile), and the flow rate, q (vehicles per hour). From the conservation of flow, we get

$$q = ks \qquad (8)$$

Vehicle speed, however, is itself a function of the vehicle density. Different empirical relations have been considered in the literature for $s(k)$. The typical flow versus density relation is shown in Figure 6.

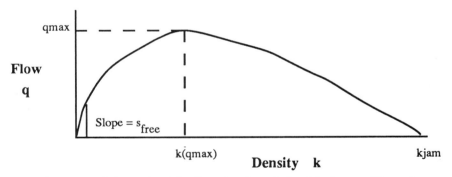

Figure 6 Schematic of the flow density relationship for a traffic path

In Table 1 we show some empirical results from [3]. These are representative data aimed at giving the reader a feel for the numbers. For our application the flow determines the rate of handoffs and location updates. The density (along with the cell size) determines the call arrival rate, the rate of time-based autonomous registrations and paging. It is interesting to note that the density of vehicles at the maximum flow condition (when handoffs and location updates are maximum) is only around 20% of the maximum possible jam density.

	Jam Density k_{jam} (veh/mi-lane)	Maximum Flow q_{max}(veh/hr-lane)	Density at Max. Flow $k(q_{max})$ (veh/mi-lane)
Highway	185-250	2000	50
Tunnel	185-250	1350	55
Arterial Street	185-250	600	40
	(person/sq.ft.)	(person/min-ft)	(person/min-ft)
Pedestrians	0.4	25	0.2

Table 1 Typical flow and density parameters

5.2 Model Parameters

We introduce an aggregate model for a city, a portion of a city or a suburban area, (say) that is covered by a single cellular switch. The roadway-density, r, is the length of the streets in lane-kilometers per square kilometer in an urban or suburban area. The parameter r, distinguishes downtown areas from sprawling urban areas from suburbs. The vehicular traffic density per square kilometer, D is a function of the roadway-density r and the vehicle density, k, per lane-kilometer $D = kr$.

The density of vehicles per lane-kilometer k can be related to the speed s and flow q of vehicles, as described in the previous section. To keep our model simple, we consider an aggregate cellular coverage area (say, a city). We apply flow density relationships in a simplistic way, so that spatial variations are incorporated in the model. Incorporating temporal variations requires a more microscopic view of the system, that we ignore for the present.

We introduce a parameter c to indicate the proportion of the coverage area that is congested (for example, in the rush-hour). The total roadway is assumed to belong to one of two regimes, (i) congested with high density and slow moving vehicles and (ii) free-flow with fast moving vehicles. Each regime has a characteristic (average) density and speed. Thus, we have divided the vehicular traffic into two components. The third component of mobile users are the pedestrians; their characteristic speed is another parameter. We choose not to use pedestrian density as a parameter of our model, at this stage. Instead, we will directly introduce the pedestrian-generated teletraffic later.

We define the mean straight-line travel as another parameter, which incorporates the propensity of vehicles and pedestrians to make a turn. With linear microcells along highways and urban streets, the handoff rate is determined by the length of the microcell and the average distance traveled by the mobile before making a turn. This is discussed in Section 5.3

We next define the parameters that relate the mobility model to the teletraffic and cellular system model. The average teletraffic generated from each vehicle

α_v (in Erlangs per vehicle) is a parameter that can projected from current cellular systems. We point out that α_v is also the in-call probability of vehicles, defined earlier in Section 4.2. Other teletraffic parameters include the average call duration τ, and the number of voice calls carried by a microcell and a macrocell base station. We will assume that the reuse factor for microcells is larger than the reuse factor for macrocells [8]. That is, with the same total number of channels, a system with microcells only gets fewer channels per cell compared to a system with macrocells only. This is quantified in Section 6.

Instead of modeling the pedestrian flow and density in detail, we directly identify the proportion of teletraffic from pedestrians, p as a parameter. This parameter varies from city to city (and season to season). By varying this parameter, we are able to study microcell and macrocell coverage areas, as a function of the pedestrian penetration of PCN's.

The model parameters are listed in Table 2, along with the nominal values used in the results presented in Section 6.

Mobility Parameters

roadway density	25	lane-km/sq-km
vehicle density (congested)	140	vehicles/lane-km
vehicle density (free-flow)	20	vehicles/lane-km
vehicle speed (congested)	5	km/hr
vehicle speed (free-flow)	25	km/hr
pedestrian speed	1	km/hr
vehicle straight-line travel (congested)	800	meters
vehicle straight line travel (free-flow)	1600	meters
pedestrian straight-line travel	160	meters
proportion of roadway congested (c)	variable	nominal value = 0.5

Teletraffic Parameters

proportion of vehicles in-call (α_v)	0.2	
proportion of pedestrian calls (p)	variable	varied between 0.0- 0.9
call duration	150	seconds
simultaneous calls per microcell	25	
simultaneous calls per macrocell	50	100 with macrocells-only

Table 2 Parameters used in this study and nominal values

5.3 Approximate Calculation of Handoff Rate per Call

If the mean sojourn time of a mobile in a cell is t and the mean call holding time is τ, then from the result of Section 4.1, equation (10), the number of handoffs per call, $E[H] = \tau/t$. Let us estimate the mean sojourn time in a cell for macrocells and microcells. We consider the mobile to be moving at a speed s through a cellular coverage area. Below we will estimate the mean travel length of the mobile in a cell, denoted by L, so that $t = L/s$, and

$$E[H] \quad = \quad \tau s / L \tag{9}$$

5.3.1 Macrocells

As a simplification we consider macrocells as squares of side a. If the mobile travels in a straight line across the cell parallel to a cell boundary, it travels exactly distance a in the cell. Further, if all traffic paths are parallel to the cell boundaries and the mobile makes exactly one 90 degree turn in the cell, we find that the average distance traveled in the cell is again a. It is unclear how generally this holds, but for the purpose of our estimates we assume that for macrocells, the mean travel distance in a cell, $L = a$. We believe that this is within a factor of two of the actual mean.

5.3.2 Microcells

Next, we consider the mobile traveling at speed s through highway or street microcells of length b. Further, assume that the mean straight-line travel, the distance that the mobile travels before making a turn, is given by d. We need to consider two distinct cases depending on whether the mobile enters the microcell at one end, or turns into the microcell somewhere in the middle.

First, consider the case when the mobile enters the microcell at *one end*. The mobile can leave the cell by either traversing length b, or by making a turn before reaching the other end. For simplicity, we assume that the mean distance between turns is exponentially distributed with mean d. If the turn occurs before the mobile traverses the length b, the mobile leaves the microcell at that distance, denoted x; otherwise, the mobile leaves the microcell after traveling a distance b. Hence, the distance traveled in the microcell, given that the mobile enters at an end, L_{end} is given by,

$$L_{end} \quad = \quad E_X[\min(x, b)] \tag{10}$$

which after some simple manipulation gives,

$$L_{end} \quad = \quad d(1 - e^{-(b/d)}) \tag{11}$$

Next, consider the case when the mobile enters the microcell *somewhere in the middle*. Then, it *either* leaves the microcell at one of the ends after traveling a distance y, where y is uniformly distributed in (0,b], *or* it makes a turn. Using the result above, we can calculate L_{mid} as follows:

$$L_{mid} \quad = \quad E_Y[d(1-e^{-(y/d)})]$$

$$= \quad d\left(1 - \frac{d}{b} + \frac{d}{b}e^{-(b/d)}\right) \tag{12}$$

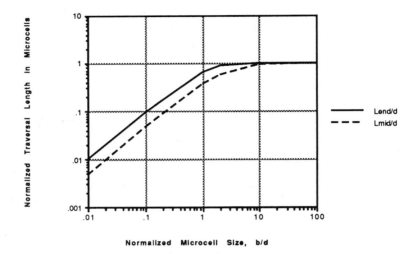

Figure 7 Mean traversal length versus microcell size normalized
by the mean straight-line travel, d

In Figure 7 we show L_{mid} and L_{end} versus the microcell size, all normalized by the mean travel length d. A mobile actually enters each microcell either at one end, or at some point in the middle, hence the *mean* traversal distance L is between L_{mid} and L_{end}. When b/d >> 1, in large microcells or for pedestrians with a high propensity to make turns in a short distance, the two travel lengths are approximately d (Figure 7). When b/d << 1, it is likely for high speed mobiles to traverse several microcells before making a turn. In that case, the probability that the mobile enters a microcell from the end is high, so that the mean travel length L, is closer to b (which is L_{end}), rather than b/2 (L_{mid}).

In the next section, we use these estimates of the mean travel length in microcells to study the handoff rate using (12). Since the handoff rate is inversely proportional to the mean travel length, the lower bound on the travel length gives an upper bound on the handoff rate, and vice versa. However, from Figure 7, we see that the bounds are within a factor of two of each other.

6. Applications
6.1 Vehicular Traffic with Macrocells only
Suppose that the pedestrian-generated teletraffic is zero, and the in-call probability for a vehicle, α_v is 0.2. This is quite high and for this study we will keep it fixed at this value. We assume that the number of channels per

macrocell is fixed at 100. Using the vehicle density and roadway density values from Table 2, the macrocell area can be calculated. As a simplification we assume square macrocells. As the congestion parameter c is increased from 0.1 to 0.9, we find that the macrocell size decreases by a factor of two from approximately 790 meters to 395 meters.

Next, we consider the handoff rate per call. For a given macrocell size, we use the result of Section 5.3.1 that the mean length of travel in a square macrocell is equal to the length of a side of the square. The handoff rate can be calculated from the call holding time and the mobile speed from Table 2, using (12). We notice two distinct trends in Figure 8. The *mean* handoff rate (for all vehicles) is actually 50% higher with the larger cell size (790 m), since in the larger cell most of the traffic is free-flowing. On the other hand, the handoff rate for both the free flowing and the congested traffic is a factor of two larger when the cell size is 395 meters. Notice also that even with the high in-call probability of 0.2, that is each vehicle generates 0.2 Erlangs of traffic, we get cell sizes of the order of 500 meters, and a mean handoff rate between 2 and 3 handoffs per call.

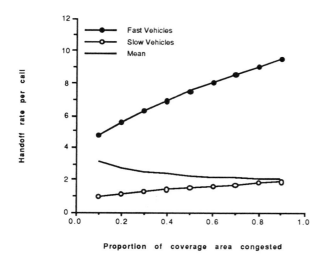

Figure 8 Handoffs per call in macrocells as a function of congestion (no pedestrian penetration). The mean rate approaches the rate for slow vehicles as the proportion of slow vehicles approaches 100%.

6.2 Microcells with Overlaid Macrocells

Let us consider an architecture of microcells overlaid with macrocells. The microcells are assumed to carry both the pedestrian traffic as well as the traffic generated from the slow-moving and congested vehicles. The traffic from fast moving vehicles is carried in the macrocells. Let us assume that the 100 channels per cell available in the macrocells-only system are divided up so that 50 channels are available in every overlaid macrocell, and 25 channels are available in every microcell. This choice is somewhat arbitrary, but reflects our assumption that no additional spectrum is used, and the reuse factor is two times larger in microcells compared to macrocells.

We fix the congestion parameter c at 0.5, that is half the coverage area is congested, and therefore fix the vehicle generated traffic. The macrocell size is fixed and found to be 1000 m. At zero pedestrian penetration, p=0.0, only the slow moving vehicles are carried in the microcells. Using the density of slow-moving vehicles from Table 2, the microcell size is found to be 1786 m. Then using the mean straight-line travel for slow vehicles (Table 2) and applying (14) and (15), we can calculate the mean length of travel for slow vehicles in the microcells. The handoff rate for slow moving vehicles is calculated using (12) and is found to be approximately 1.5 per call.

As the pedestrian penetration increases, the handoff rates for pedestrians and vehicles are similarly calculated using parameter values from Table 2 in equations (14), (15) and (12). The results are shown in Figures 9 and 10, respectively. The lower and upper bounds on the handoff rates as discussed in Section 5.3.2 are plotted. As discussed in that section, the bounds are close for large values of b/d. This can be seen in Figure 9, for small p.

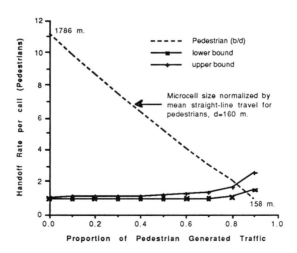

Figure 9 Handoff Rate per call for Pedestrians as a function of pedestrian penetration

In Figure 10, and for large p in Figure 9, notice that the bounds are within a factor of 2 of each other. This was shown in Section 5.3.2, where we pointed out that for small values of b/d, the actual handoff rate is expected to be closer to the lower bound (as the mean travel distance is closer to its upper bound). This applies to the large values of p in Figure 10, so that the handoff rate for slow vehicles increases slower than that suggested by the upper bound.

The mean handoff rate, averaged over the pedestrian, slow vehicles and fast vehicles, is between 1.3 and 2.0 handoffs per call as long as the pedestrian generated traffic is less than 60% of the total. Beyond that the mean rate increases slowly.

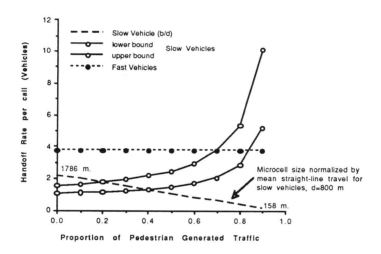

Figure 10 Handoff Rates for Vehicles as a function of pedestrian penetration

7. Conclusions

Many results in this paper have been stated as "folk-truths" by various researchers in the area. Here, we have provided conditions under which some of these oft-stated results hold. The proofs offer new insight into the problem of teletraffic modeling for PCN's. In this category are results such as: (i) The handoff rate increases as the square root of the increase in the cells per unit area (Section 2); (ii) The handoff rate is minimized if cells are shaped so that the longer boundaries are along direction of larger flow (Section 3); (iii) The handoff rate is given by the ratio of mean call duration to the mean cell sojourn time

(Section 4.1); and (iv) The actual handoff rates do not increase rapidly at high user density (high pedestrian penetration) due to the decrease in user mobility (Section 6).

In the new results category, we have shown that: (i) The handoff rate increases *linearly* as new boundaries are introduced along traffic paths (Section 4.2). We discussed that this result is consistent with the widely-believed *square-root* result of Section 2, when the cell is large and is not dominated by a single or a few traffic paths. In microcells the handoff rate increase is linear with increasing cells per unit area; and (ii) We derived expressions for the mean length of travel in microcells using the microcell size, and the mean straight-line travel distance of the mobile, that is, the distance the mobile travels before making a turn (Section 5.3). For pedestrians, in microcells that are *much larger* than the mean straight-line travel, the handoff rate is dominated by the pedestrian making a turn out of the microcell. Hence, it is approximately constant until the microcell size decreases to about twice the mean straight-line travel. On the other hand, for fast-moving cars in small microcells the handoff rate is determined by the microcell size, so that the handoff rate increase is linear with increasing mobile density. Since fast-moving vehicles experience a disproportionately large number of handoffs, by confining them to large overlaid macrocells we decrease their handoff rate. Moreover, by confining them to macrocells, we also restrict their handoff rate increase to the square-root of the increase in the density of mobiles in the overlaid macrocells.

The parametric model offers a quantitative way of studying design tradeoffs. The model cannot be used at this point to study resource allocation, handoff algorithms or other studies that require detailed modeling at the individual cell level. Although that is our goal, so far we have a model of an aggregate cellular system that is useful to study design issues concerning cell and switch coverage areas, microcell versus macrocell tradeoffs etc. The actual numbers here are highly dependent on the chosen parameter values (Table 2), however, the model offers an invaluable tool to study trends.

The microscopic approach in the WINLAB simulator [5] permits the flow-density dynamics of interacting street blocks to be incorporated in the model. The WINLAB simulator attempts to model real traffic scenarios, exactly. However, it is easier to study gross trends using a simpler abstract model discussed here. The WINLAB simulator requires verification with real traffic data collected from cities; we aim to verify our model against experiments on the WINLAB simulator.

Acknowledgment

Discussions with Robert Morris, Bharat Doshi, Wing Wong, Behrokh Samadi and Rodolfo Milito, are gratefully acknowledged. The exposition has benefited from the comments of a number of colleagues. I thank them for their careful reading and detailed comments. This version of the paper is much improved due to their efforts.

References

[1] S. Nanda and D. J. Goodman, "Dynamic Resource Acquisition: Distributed carrier allocation in TDMA cellular systems," in **Third Generation Wireless Information Networks** (Proceedings of the Second WINLAB Workshop, Oct. 1990), Kluwer: Norwell MA, 1992.

[2] E. Alonso, K. S. Meier-Hellstern and D. R. O'Neil, "The use of SS7 and GSM to support high density personal communications," *WINLAB Technical Report TR-24*, December 1991.

[3] A. D. May, **Traffic Flow Fundamentals**, Prentice Hall: Englewood Cliffs NJ, 1990.

[4] S. A. El-Dolil, W.-C. Wong and R. Steele, "Teletraffic performance of highway microcells with overlay macrocell," *IEEE Journal on Selected Areas in Comm.*, vol. J-SAC 7, No. 1, pp. 71-78, January 1989.

[5] I. Seskar, S. V. Maric, J. Holtzman and J. Wasserman, "Rate of location area updates in cellular systems," *Proc. IEEE Vehicular Technology Conference*, Denver, May 1992.

[6] L. Kleinrock, **Queueing Systems Volume 1: Theory**, Wiley: New York, 1975.

[7] N. Amitay, L. J. Greenstein and G. J. Owens, "Measurement-based estimates of BER performance in urban LOS microcells at 900 MHz," *Proc. IEEE Veh. Tech. Conf.*, pp. 904-909, May 1991.

COMMUNICATIONS TRAFFIC PERFORMANCE
FOR CELLULAR SYSTEMS
WITH MIXED PLATFORM TYPES

Stephen S. Rappaport

Department of Electrical Engineering
State University of New York
Stony Brook, New York 11794-2350

ABSTRACT

Cellular communication systems that support a mixture of platform types distinguished by different mobility characteristics are considered. A tractable analytical model for traffic performance analysis is developed using multi-dimensional birth-death processes and the method of phases. The framework allows consideration of homogeneous and non-homogeneous systems, a broad class of dwell time distributions, and "missed" hand-off initiations. Cut-off priority for hand-offs and several platform types are considered to demonstrate the approach. The effects of different mobility parameters and of imperfect detection of hand-off needs are examined. Theoretical performance characteristics are obtained. These exhibit carried traffic, hand-off activity, blocking probability and forced termination probability for each platform type. The realizable exchange of blocking for hand-off performance is shown.

The author is grateful to Mr. Gustavo MONTE for skillful computer programming that generated the performance characteristics shown.

The research reported in this paper was supported in part by the U.S. National Science Foundation under Grant No. NCR-9025131 and in part by IST/SDIO under Grant No. N00014-91-J-4063 administered by the U.S. Office of Naval Research.

Some of the computations were obtained using the Cornell National Supercomputer Facility, a resource of the Center for Theory and Simulations in Science and Engineering (Cornell Theory Center), which receives major funding from the U.S. National Science Foundation and IBM Corporation, with additional support from New York State and members of the Corporate Research Institute.

INTRODUCTION

The hand-off problem arises in cellular mobile and personal communication networks when a communicating platform moves from a spatial (source) region served by one wireless gateway to a (target) region served by another. When circuit or virtual circuit switching is used, continuation of a call depends on several factors. These include: {1} reliable detection of conditions that indicate the need of a platform for a hand-off; {2} identification of alternative gateway(s) for service; {3} timely exchange of supervisory signals; and, {4} deployment and availability of communications resources. Missed or unsuccessful hand-offs result in forced termination of calls and are perceived by users as interruptions in service. As smaller cell sizes are used to increase overall system capacity, more hand-offs are needed to sustain calls to satisfactory completion. Thus the problem is becoming increasingly significant as mobile, portable and personal communications proliferate and small-cell high-capacity systems are devised.

Early studies of hand-off issues relied exclusively on computer simulation [1]. Development of realistic analytically tractable models has been the thrust of some more recent work [2]-[8]. Among these is a methodology, which is based on the notion of multi-dimensional birth-death processes [6]-[8]. It permits the computation of theoretical performance characteristics, provides considerable additional insight into the problem, and alleviates the sole dependence on simulation techniques - even for situations involving a fair amount of physical complexity. The method depends heavily on the assumption of memoryless properties for various driving processes [9]. Similar assumptions in other contexts have served telecommunications traffic theory well for many years, but since less is known about some of the driving processes that arise in the cellular communications context, it is important to extend the applicability of the approach.

This paper extends our earlier results and mathematical framework in several important directions. Firstly, we present a state variable description that allows consideration of a broad class of dwell time distributions which can accommodate empirical data, while at the same time retaining the conditions that allow an analytically tractable solution. Secondly, a mixture of platform types (with widely differing mobility characteristics) in the same system is considered. Thirdly, the analytical model is extended to include the effects of imperfect hand-off initiation. Specifically we develop formulas that relate overall hand-off performance to two component problems - *hand-off initiation* and *resource availability*. Finally, we present example results. These were generated using (when appropriate) a stand-alone workstation or remote access to supercomputing facilities.

In previous work we used the concept of *dwell time* to characterize the amount of time that a mobile platform is *within communications range* of a given gateway. Figure 0.1 shows an *approximate* relationship between cell sizes and dwell times for various platform types. Estimated speeds of the platform types listed are very rough, and of course, there is no definitive agreement on terminology for cell sizes. The figure is NOT intended to suggest, for example, that jetliners and pedestrians should be supported by the same system and on the same channels. Indeed, strong arguments can be made which suggest that even hand-held and automobile mounted units should be serviced by *different* wireless systems with appropriate interconnecting gateways [10],[11],[12]. But the figure DOES illustrate, that for a given cell size and system, a variety of platform types (with a range of mobility characteristics) should be supported. Moreover, there exists for example, the possibility that hand-held units (intended for pedestrians) will frequently be used on more highly mobile platforms such as automobiles. Such use can have a significant impact on communications traffic performance. Thus the consideration of mixed platform types is an issue that should be considered. The analytical basis for Figure 0.1 is given in [13].

Clearly, dwell time depends on many factors such as propagation conditions, the path a mobile platform follows, its velocity profile along the path, and especially the definition of *communications range*. But even if all of these were known, the dependency is so complex and burdensome that one would eventually have to resort to empirical findings in some way. Much of the difficulty can be circumvented by characterizing dwell time as a random variable whose distribution function is known but chosen to accommodate empirical data.

The approach used is to consider the dwell time to be a random variable that is the sum of independent negative exponential random variables. Then by using an appropriate definition of state variables, the memoryless property can be maintained, and a multi-dimensional birth-death process model can be used. Dwell times are considered to consist of several phases. This artifice permits consideration of a broad class of probability density functions (that are not memoryless) within the framework. A similar generalization for unencumbered session times is also possible but is not developed here for two reasons. Firstly, the approach tends to increase the number of states needed to represent the system. Secondly, *session* duration times that have negative exponential distributions are not uncommon in telecommunications traffic theory, so there are ample precedents. We focus on dwell time characteristics, because much less is known about these, and as explained above it seems that a broad class of distributions that characterize dwell times would be useful in modeling actual cellular systems.

Theoretical performance characteristics are calculated and presented. These show carried traffic, hand-off activity, blocking probability and forced termination probability for each platform type. The characteristics exhibit the realizable exchange of blocking for hand-off performance. The effects of different platform mobility parameters on performance are investigated. *Extended performance measures* which account for: {1} the ability of the detection/decision algorithm to initiate the hand-off procedure in sufficient time to allow the required exchange of supervisory signals, as well as, {2} the availability of communications resources in the target cell, are developed. The approach and results are applicable to cellular and micro-cellular systems. For some parameter choices, calculations to generate the results were done on readily available workstations. For others, where a large number of states were required, calculations were done using remote access to the Cornell National Supercomputer Facility.

MODEL DESCRIPTION

The development described here proceeds along the lines of [6]-[8]. For brevity, we proceed quickly to a mathematical statement of an example problem. A more complete discussion is given in [13]. In a subsequent section we consider the effects of failure to initiate the hand-off procedure when in fact a hand-off is necessary. For this purpose one has to distinguish between a hand-off "need" and a hand-off attempt. We consider an *attempt* to arise only when the hand-off initiation procedure is (correctly) activated for a *valid need*. We begin however, by first assuming that the hand-off "need" detection and initiation procedure is perfect. That is all valid needs are detected and no invalid needs activate the hand-off procedure. The results obtained are then extended.

Example Problem Statement - Single Call Hand-Offs, Cut-Off Priority, Mixed Platform Types, General Mobility Characteristics.

There are G types of mobile platforms, indexed by $g = 1, 2, \ldots G$. No platform can support more than one call at any given time.

The new call origination rate *from a non-communicating g-type platform* is denoted $\Lambda(g)$. We define $\alpha(g) = \Lambda(g) / \Lambda(1)$.

The number of *non-communicating* g-type platforms in any cell is denoted $v(g,0)$. The total rate at which new g-type calls are generated *in a cell* is denoted $\Lambda_n(g)$. Thus, $\Lambda_n(g) = \Lambda(g) \cdot v(g,0)$.

CHANNEL LIMIT: Each cell or gateway can accommodate C channels.

CHANNEL QUOTAS: At any gateway, the maximum number of channels that can be simultaneously used by g-type platforms is $J(g)$.

CUT-OFF PRIORITY: C_h channels in each cell are reserved for hand-off calls. New calls will be blocked if the number of channels in use is $C - C_h$ or greater. Hand-off attempts will fail if the number of channels in use is C.

Note: It is assumed that $v(g,0) >> C$, so that overall the population of non-communicating g-type platforms in a cell generates $\Lambda_n(g)$ calls per second. This *infinite* population model is consistent with a large population of non-communicating g-type platforms in each cell, only a small fraction of which are served at any time. This is, in fact, usually the case.

The unencumbered call (session) duration on a g-type platform is a random variable, $T(g)$, that has a negative exponential distribution (ned). The mean of the random variable is $\overline{T}(g) = 1/\mu(g)$.

The dwell time in a cell for a g-type platform is a random variable, $T_D(g)$ having a mean $\overline{T}_D(g)$. The random variable, $T_D(g)$ is the sum of $N(g)$ statistically independent ned random variables denoted $T_D(g,i)$, where $i = 1,2,3,\ldots N(g)$. The mean of $T_D(g,i)$ is $\overline{T}_D(g,i) = 1/\mu_D(g,i)$ and its variance is $VAR(T_D(g,i) = 1/[\mu_D(g,i)]^2$. Thus

$$\overline{T}_D(g) = \sum_{i=1}^{N(g)} \overline{T}_D(g,i) = \sum_{i=1}^{N(g)} 1/\mu_D(g,i) \qquad , \qquad (1)$$

and

$$VAR[T_D(g)] = \sum_{i=1}^{N(g)} 1/[\mu_D(g,i)]^2 \qquad . \qquad (2)$$

The *squared coefficient of variation* for the dwell time of a g-type platform is the ratio of its variance to the square of its mean. This is given by

$$\kappa^2[T_D(g)] = VAR[T_D(g)] / [\, \overline{T}_D(g) \,]^2 . \qquad (3)$$

An interesting special case occurs if $\mu_D(g,i)$ does not depend on i. Then $T_D(g)$ has an Erlang distribution with a mean $N(g)/\mu_D(g)$, and a squared coefficient of variation, $1/N(g)$. For each platform type, the two parameters can be easily chosen to accommodate empirical data by using the mean and variance of observed dwell times. A value of $\kappa^2=1$ corresponds to a ned random variable, while $\kappa^2=0$ corresponds to the deterministic case (constant dwell time).

The problem is to calculate relevant performance characteristics, including, (for each platform type): Blocking Probability, Hand-off Failure Probability, Forced Termination Probability, Carried Traffic, and Hand-Off Activity Factor.

Notes: We consider *blocking probability* to be the average fraction of **new** call originations that are denied access to a channel. *Hand-off failure probability* is the average fraction of hand-off "needs" that fail to gain access to a channel in the target zone. *Forced termination probability* is the probability that a call will suffer a hand-off *attempt* failure some time in the "lifetime" of the call. *Hand-off activity factor* is the average number of hand-off attempts for a call that receives service.

We can formulate a solution to this problem using the framework described in [6] - [8]. The major differences are in definition of the state variables, identification of the driving processes, and determination of the state transition probability flows. In what follows we emphasize those aspects of the mathematical development that differ. We consider that when a platform of type g enters a cell, it passes through $N(g)$ phases of dwell time. These are identified by the index, $i=1,2,3,\ldots N(g)$. The amount of time spent in each phase is $T_D(g,i)$, a ned random variable as defined above.

STATE CHARACTERIZATION

First consider a *single* cell. We define the state (of a cell) by a

182

sequence of non-negative integers. These can be conveniently written as G n-tuples.

$$
\begin{array}{l}
v_{11}, v_{12}, v_{13}, \cdots v_{1N(1)} \\
v_{21}, v_{22}, v_{23}, \cdots\cdots\cdots\cdots v_{2N(2)} \\
\vdots \quad \vdots \quad \vdots \qquad\qquad \vdots \\
v_{g1}, v_{g2}, v_{g3}, \cdots\cdots\cdots\cdots\cdots v_{gN(g)} \\
\vdots \quad \vdots \quad \vdots \qquad\qquad\quad \vdots \\
v_{G1}, v_{G2}, v_{G3}, \cdots\cdots\cdots\cdots v_{GN(G)}
\end{array}
\tag{4}
$$

where v_{gi} { $g=1,2,...G$; $i=1,2,...N(g)$ } is the number of platforms of type g that are in phase i. It was found convenient to order the states using an index $s=0,1,2,...s_{max}$. Then the state variables v_{gi}, can be shown explicitly dependent on the state. That is, $v_{gi} = v(s,g,i)$.

When the cell (gateway) is in state, s, the following characteristics can be determined:

The number of channels being used by g-type platforms is

$$
j(s,g) = \sum_{i=1}^{N(g)} v(s,g,i) \qquad .
\tag{5}
$$

The total number of channels in use is

$$
j(s) = \sum_{g=1}^{G} j(s,g) \qquad .
\tag{6}
$$

Permissible states correspond to those sequences for which all (limit and quota) constraints are met. Additional constraints can also be considered within this same framework. Here we have a *channel limit* which requires, $j(s) \leq C$, and *channel quotas* which require, $j(s,g) \leq J(g)$, for $g=1,2,...N(g)$.

A thorough formulation which accounts for direct coupling of **cell state** transitions of adjacent cells that are involved in a hand-off is circumvented by relating the average hand-off arrival and departure rates as in [6] and [7]. This avoids having to deal with an enormously (and usually intractably) larger number of **system states** represented by sequences of all simultaneously possible **cell states**. Both *homogeneous* and *non-homogeneous* cellular systems can be treated in this way. The result is that we only have to consider a **single cell** and deal with the number of states needed to characterize its behavior. Even with this simplification, the number of states that are needed can be quite formidable for certain parameter choices. So, it is important that the procedures and algorithms used to solve the resulting equations be chosen, devised, and organized so that computational solutions are feasible. The number of cell states needed to represent a system with two platform types each having two phases of dwell time and no channel quotas is shown in Table 1.

TABLE 1: NUMBER OF CELL STATES NEEDED
$G=2$, $N(1)=2$, $N(2)=2$, CHANNEL LIMIT $= C$.

C	number of states
10	1,001
15	3,876
20	10,626
25	23,751
30	46,376

DETERMINATION OF THE EQUILIBRIUM STATE PROBABILITIES

For this example there are five relevant driving processes. These are: $\{n\}$ The generation of _new_ calls in the cell of interest; $\{c\}$ The _completion_ of calls in the cell of interest; $\{h\}$ The arrival of communicating vehicles at the cell of interest; $\{d\}$ The _departure_ of communicating vehicles from the cell of interest; and, $\{\phi\}$ The transition between dwell time phases. Because there are different platform types, all of these processes are multi-dimensional. As before we use Markovian assumptions for the driving processes in order to render the problem amenable to solution using multi-dimensional birth-death processes. Specifically, in addition to the previous assumptions we assume that: $\{1\}$ The new call arrival processes in any state are Poisson point processes with state dependent means; and that, $\{2\}$ The hand-off call arrivals for each platform type are Poisson point processes.

Let Λ_h and Δ_h respectively denote the average hand-off arrival and departure rates for a cell. Also let F_g and F'_g respectively denote the fraction of arrival and departure attempts of type g platforms. We note that any hand-off departure of a g-type platform from a cell corresponds to a hand-off arrival of (the very same) g-type platform to some other cell. Thus, it must be true that for a homogeneous system in statistical equilibrium, the hand-off arrival and departure rates per cell must be equal. That is we must have

$$F_g = F'_g \qquad , \qquad (7)$$

184

and

$$\Lambda_h = \Delta_h \qquad . \qquad (8)$$

For non-homogeneous cellular systems, the same basic approach can be used except that the arrival and departure rates are, in general, unequal. Instead, they are related by some (non-unity) constant that specifies the "tilt" or degree of inhomogeneity for the cell under consideration. That is, we require

$$\Lambda_h = \Theta \cdot \Delta_h \quad , \qquad (9)$$

in which, $\Theta \neq 1$, is a given "tilting parameter." The parameter is chosen to be greater or less than unity to characterize a cell which on the average experiences more or less hand-off arrivals than departures, respectively. These *conservation rules* allow the system to be characterized by considering state transitions of only **one cell**.

It is necessary to characterize all of the state transitions. For each state, the possible predecessor states must be identified. That is, those states which could have *immediately* given rise to the current state under each of the (multi-dimensional) driving processes and the state transition probability flows, must be found. The flow balance equations can then be determined and the state probabilities can be calculated. A complete discussion is given in [13]. An important consideration is that because of the *conservation rules* that must be satisfied, the coefficients that appear in the state probability flow balance equations, depend on the unknown state probabilities, p(s). Therefore the probability flow balance equations are actually a set of simultaneous nonlinear equations. However, by beginning with guesses for the average hand-off arrival rate Λ_h, and the fractions, F_g of hand-offs that arise from platforms of type g, $(g = 1,2,...G)$, the iterative approach of [6] can again be used. In this approach the system is linear within each iteration.

Because solving for the state probabilites involves nested iterations, fairly intensive computations can be required to produce results if the number of states is large. Nevertheless we were able to generate numerical performance characteristics for many non-trivial parameter choices using readily available workstations. For some additional parameter choices, calculations were done via remote access to the Cornell National Supercomputing Facility.

Some examples of cpu time requirements needed to obtain an equilibrium solution are given in Table 2. Computation of the first point in a run is more time consuming because the computer program must first identify and order the states and their predecessors. This part of the calculation need not be repeated for subsequent points. The required cpu time for a point also depends on particular parameter values. These effects produce the spread shown in Table 2.

TABLE 2: APPROXIMATE REQUIRED COMPUTATION TIME
G=2, N(1)=2, N(2)=2, CHANNEL LIMIT = C.

		CPU TIME IN MINUTES			
		Sun SparcStation SLC		Supercomputer	
C	number of states	1st pt.	other pts.	1st pt.	other pts.
15	3,876	1.32 - 6.73	1.03 - 5.34	0.43 - 2.26	0.26 - 1.72
25	23,751	-	-	7.96 - 26.65	2.33 - 16.61

PERFORMANCE MEASURES

When the statistical equilibrium state probabilities and transition flows are found the required performance measures can be calculated.

Carried Traffic
The carried traffic per cell for each platform type is the average number of channels occupied by the calls from the given platform type. If users pay for "air time" this indicates the revenue that the service provider can expect from each type of platform. The carried traffic for g-type platforms is

$$A_c(g) = \sum_{s=0}^{s_{max}} j(s,g) \cdot p(s) \qquad , \qquad (10)$$

and the total carried traffic is

$$A_c = \sum_{g=1}^{G} A_c(g) \qquad . \qquad (11)$$

Blocking Probability
The blocking probability for a call from a g-type platform is the average fraction of new g-type calls that are denied access to a channel. Blocking of new g-type call occurs if there are no channels to serve the call or if the number of g-type calls in progress is at the quota level. We define the following disjoint sets of states:

$$B_0 = \{ s: C-C_h \leq j(s) \leq C \}$$
$$B_g = \{ s: j(s) < C-C_h , j(s,g) = J(g) \} \qquad , \qquad (12)$$

in which, g= 1,2,...G .

Then the blocking probability for g-type calls is

$$P_B(g) = \sum_{s \in B_0} p(s) + \sum_{s \in B_g} p(s) \qquad . \qquad (13)$$

If there are no channel quotas, then $J(g) \geq C$, so there are no states in any of the sets, B_g, $g = 1,2,3\ldots G$. Blocking probability is then the same for any new call regardless of the platform type on which it originates.

Hand-Off Failure Probability

The hand-off failure probability for g-type calls is the average fraction of g-type hand-off attempts that are denied a channel. We note that hand-off attempts have potential access to all channels of a cell without regard to C_h, but may be subject to channel quota constraints. We define the following disjoint sets of states, in which at least some hand-off attempts
will fail:

$$H_0 = \{ s: j(s) = C \}$$

$$(14)$$

$$H_g = \{ s: j(s) < C, \ j(s,g) = J(g) \} \qquad ,$$

in which $g = 1,2,3,\ldots G$.

Then the hand-off failure probability for a g-type hand-off can be written as

$$P_H(g) = \sum_{s \in H_0} p(s) + \sum_{s \in H_g} p(s) \qquad . \qquad (15)$$

Forced Termination Probability

Perhaps more important than hand-off failure probability, is the *forced termination probability*, $P_{FT}(g)$. This is defined as the probability that *a g-type call that is not blocked is interrupted due to hand-off failure during its lifetime.* For convenience we limit our discussion here to the case where all dwell-time phases of a given platform type are statistically identical. That is, $\mu_D(g,i) \equiv \mu_D(g)$. Then $\rho_n(g,i) = 1/N(g)$. Let $\pi(g)$ be the probability that a platform of type g completes its current dwell-time phase before its call is (satisfactorily) completed. Then since dwell time phases are ned random variables, we have

$$\pi(g) = \mu_D(g) \ / \ [\ \mu(g) + \mu_D(g) \ ] \qquad . \qquad (16)$$

For a call that is being served on a g-type platform in phase i, $N(g)-i+1$ dwell time phases must be completed by the supporting platform for a hand-off attempt to be generated. Therefore the probability that such a call requires a hand-off is

$$b(g,i) = [\pi(g)]^{N(g)-i+1} \qquad . \qquad (17)$$

The probability that a new call (which is not blocked) on a g-type platform requires a hand-off is therefore

$$b(g) = \sum_{i=1}^{N(g)} \rho_n(g,i) \cdot b(g,i) \qquad . \qquad (18)$$

All calls that are successfully handed off renew service (in the target cell) in the first dwell time phase. So the probability that a call that has been handed off requires yet another hand-off is $b(g,1)$.

The probability that the call is forced to terminate on its kth hand-off attempt is

$$Y(g,k) = b(g) \cdot P_H(g) \cdot \{ b(g,1) \cdot [1\text{-}P_H(g)] \}^{k-1} . \tag{19}$$

The forced termination probability is therefore

$$P_{FT}(g) = \sum_{k=1}^{\infty} Y(g,k) \qquad . \tag{20}$$

This can be compactly written in closed form as

$$P_{FT}(g) = b(g) \cdot P_H(g) \, / \, [\, 1 - \psi(g) \,] \qquad , \tag{21}$$

in which we have let

$$\psi(g) \equiv b(g,1) \cdot [1\text{-}P_H(g)] \qquad . \tag{22}$$

Hand-Off Activity Factor
We define the hand-off activity factor, $\eta(g)$ as the expected number of hand-off attempts for a non-blocked call on a g-type platform.

In [13] it is shown that this performance measure can be compactly written in closed form as

$$\eta(g) = b(g) \cdot \varphi(g) \, / \, [1\text{-}\psi(g)]^2 \equiv b(g)/[1\text{-}\psi(g)] \qquad . \tag{23}$$

DETECTION OF HAND-OFF NEEDS:
EXTENDED PERFORMANCE MEASURES

An important aspect of the hand-off process is the detection of the need of a communicating platform for a hand-off. Up to this point, this issue has been set aside - only the limitations imposed by the availability of communication resources in the target cell have been considered. Many alternative methods for *hand-off need detection* can be devised [14]-[17]. Consideration of such methods is not the subject of this paper. Here it is assumed that *some* algorithm is used to initiate the hand-off procedure. For the present purpose, the algorithm is characterized by a parameter, ζ, which denotes the probability that a need for a hand-off is *missed*. We consider a hand-off need to be *missed* if: the conditions necessitating a platform hand-off are either not detected; or, are not detected early enough to permit the necessary exchange of supervisory signals before the current link fails. Of course, there is also a non-zero probability that a hand-off request is initiated when in fact a hand-off is NOT needed. Such requests cause unnecessary system *churning* but do not result in forced terminations. This is because even if the request cannot be accommodated in the target cell, the call can be continued in the platform's current cell. Therefore *false requests* are not considered here. We suggest that hand-off initiation schemes consider the trade-off between *missed hand-off needs* and *system churning* and characterize the former by a probability, ζ, as described above.

Because of *missed hand-off needs*, only the fraction, $1\text{-}\zeta$, of actual hand-off departures will result in the initation of hand-off attempts. Thus for a *homogeneous system*, the average hand-off attempt and hand-off departure rates (per cell) will not necessarily be equal - rather they will be related by $\Lambda_h = (1-\zeta)\cdot\Delta_h$. For the more general (homogeneous or) non-homogeneous case the relationship is

$$\Lambda_h = \Theta \cdot (1-\zeta) \cdot \Delta_h \qquad . \tag{24}$$

A *need* for hand-off will result in a forced termination if either the detection of that need is missed, or the need is detected but resources to service it are unavailable in the target cell. Thus the probability that a needed hand-off results in a forced termination is

$$P_H^*(g) = \zeta + (1-\zeta) \cdot P_H(g) \qquad . \tag{25}$$

The expressions for overall performance measures for the more general case are similar to those described in the previous sections except that $P_H(g)$ in equations (21)-(23) should be replaced by $P_H^*(g)$ as given by (25).

Given a *hand-off need detection scheme* for which ζ is known, the calculation for the more general case proceeds by solving the equilibrium state probability flow balance equations using the *conservation rule* (24). The resulting state probabilities are then used to determine the carried traffic components and blocking probabilities using (10)-(13). The hand-off need failure probabilities, $P_H(g)$, are calculated using (15) and (25). The forced termination probabilities are calculated using

$$P_{FT}^*(g) = b(g) \cdot P_H^*(g) \diagup [1 - \psi^*(g)] \qquad , \tag{26}$$

in which

$$\psi^*(g) = b(g,1) \cdot [1-P_H^*(g)] \qquad . \tag{27}$$

In this case, the hand-off activity factor for a g-type call is the expected number of NEEDED hand-off attempts generated by a non-blocked call on a g-type platform. This is given by

$$\eta^*(g) = b(g) \diagup [1 - \psi^*(g)] \qquad . \tag{28}$$

Overall traffic performance characteristics that account for the reliability of the hand-off initiation algorithm as well as the limitations due to communications resource availability and organization were calculated using these formulas.

DISCUSSION OF RESULTS

For the purpose of obtaining some example numerical results the homogeneous case ($\Theta=1$) with *perfect* hand-off need detection ($\zeta=0$) was used for Figures 1.X, 2.X and 3.X. Figures 4.X are for the homogeneous case with *imperfect* hand-off need detection. Figures 3.X are for a single user type. Figures 1.X and 4.X are for two platform types. Additional figures given in [13] are for a situation with three platform types; high mobility, low mobility, and stationary. For all figures an unencumbered call duration of 100 sec. was assumed and only a channel limit, C, (no quotas), was considered. In Figures 1.X, 3.X, and 4.X, C=15 was used. Figures 2.X are for C=25. The figures presented are of two kinds. One kind has an abscissa reflecting call demand (with dwell times held fixed). In these the abscissa is new call origination rate for platform type 1 (denoted $\Lambda(1)$) with the *ratio* of new call origination rates from other platform types held fixed with respect to type 1. The other kind has an abscissa reflecting platform dwell times (with new call origination rates held fixed) and with dwell times of other platform types held in fixed *ratio* to that of type 1. Because of the scaling for these figures, the abscissa can also be envisioned as proportional to cell size.

For Figures 1.X and 4.X, two platform types (G=2) were considered, a low mobility platform type and a high mobility platform type. The mean dwell times of these platform types were taken in the ratio of 5 to 1, throughout. For an appropriate cell size, these choices can represent pedestrians and autos, respectively. Figures 3.X, which are for a single platform type, show the effects of different dwell time distributions (all having the same mean but with different coefficients of variation). For convenient reference, the specific parameters used for each figure are included in Table 3, at the end of this section.

Figure 1.1 shows blocking and forced termination probabilities as a function of new call arrival rate from type 1 platforms. For the chosen parameters the abscissa range corresponds to a combined new call arrival rate *per cell* ranging from 1.8 to 8.10 calls per minute. Since there are no channel quotas, the blocking probability is the same for each platform type. However since the platforms have different (mobility parameters) mean dwell times, there are differences in forced termination probability. Increasing the value of C_h, reduces forced termination probability at the cost of increasing blocking probability. For example, at an abscissa value corresponding to (a type 1 new call arrival rate) $\Lambda(1)=2.75E-04$, the blocking probability changes from 1.02E-02 to 3.71E-02 (a factor of 3.64) as C_h is increased from 0 to 2, while forced termination probabilities for platforms of both types are *each* decreased by a factor of approximately 20.8 (from 1.03E-03 to 4.90E-05 for type 1 platforms, and from 5.08E-03 to 2.45E-04 for type 2 platforms.

Figure 1.2 shows the total carried traffic per cell and the traffic components from each platform type. For low demand, the carried traffic increases linearly with increasing demand. For higher demand the increase in traffic is less than the proportional increase in demand. This is especially true for large C_h since blocking performance is sacrificed to accommodate hand-offs. As an example, for $\Lambda(1)=2.75E-04$ calls/sec., the total carried traffic is 8.05 erlangs for $C_h=0$, but only 7.86 erlangs for $C_h=2$. From Figure 1.4, it is seen that this decrease in carried traffic depends on the ratio $\delta = \overline{T}(g) / \overline{T}_D(g)$. As C_h increases, there are compensating trends. For values of $\delta > > 1$, there tends to be many hand-off attempts. So the trend toward a decrease in carried traffic due to increased C_h (and increased blocking) is offset by a reduction in forced termination probability, which allows more calls to be carried to successful completion.

Figures 1.3, 1.4, and 1.5 show dependence of performance on dwell time for a type 1 platform with the ratio of dwell times held fixed at 5 to 1. If one considers dwell time to be proportional to cell size, these figures can be interpreted as showing dependence of the performance measures on cell size (but with demand in a cell being held fixed). In these figures the value of $\Lambda(1) = 2.75E-04$ calls/sec. was assumed. In Figure 1.3 we see that for small values of the abscissa, blocking probability increases with increasing dwell time, and from Figure 1.4 carried traffic increases as well. For forced termination probabilities, there are two opposite effects. Increasing the dwell time tends to reduce hand-off attempts and therefore to reduce forced termination probability. But then more calls are sustained to completion so the cells carry more traffic, and this tends to increase forced termination probability. The net effect is an increase in the forced termination probability for small dwell times (cell size). With further increasing cell size, the carried traffic tends to saturate, but hand-off activity continues to decrease. Thus forced termination probability decreases.

Figure 1.4 shows the carried traffic per cell. It is seen that for small cell size, $(\delta > > 1)$ most of the total carried traffic is from the less mobile platform type. This is because the high mobility platforms (since they require more hand-offs) are more likely to encounter a hand-off failure. For large cell sizes ($\delta < < 1$), there is in the limit no impact due to different mobility characteristics, because all calls will be completed in the cell where they originated. Thus (with equal new call origination rates) the platform types are equally likely to get and retain a channel. The carried traffic of each platform type tends to the same value.

Figure 1.5 shows hand-off activity of each platform type. For the parameter choices shown, this was found to be essentially independent of C_h, but strongly dependent on cell size. This is because in the usual range of interest, $P_H < < 1$, so essentially all hand-off attempts succeed and the average number of hand-off attempts is determined mainly by mobility parameters.

A completely analogous discussion to that given above for Figures 1.X (for which C=15) applies to Figures 2.X (for which C=25).

In Figures 3.1 and 3.2 the effects of different dwell time probability density functions (pdf's) are considered. For Figures 3.X, there is only one platform type. For a given abscissa in Figure 3.1, the mean dwell time is held fixed and blocking and forced termination probabilities are shown for values of N(1)=1,2,3,4. These values correspond to squared coefficients of variation of 1, 0.5, 0.333, and 0.250, respectively. The number of states needed to characterize performance increases rapidly with N(1). For the parameters of Figures 3.X, the number of states required for N(1)=1,2,3 and 4 is 16, 136, 816 and 3876, respectively. Increasing N(1) corresponds to dwell times tending more toward the deterministic case. It is seen that both blocking and forced termination probabilities tend to increase as dwell time becomes less random. The effect of different numbers of phases is most noticeable for small cell size. Unless the dwell time is somewhat less than the unencumbered session holding time, the predominant parameter that affects the performance characteristics is mean holding time. *This fortuitous result allows one often to calculate relevant performance characteristics using the negative exponential model for dwell times.* This permits computation using the smallest number of states within this framework.

Figure 3.2 shows hand-off activity for C_h=0,2,4, and N(1)=1,2,3. The mean dwell time was taken to be 20 sec. At low demand the hand-off activity is close to 100/2 = 5. This is the ratio of unencumbered session duration to mean dwell time and roughly the expected number of cell boundary crossings for a call that is sustained to completion. This agrees with the mathematical limits developed in [13]. As demand increases, the hand-off activity tends to decrease because hand-off failure becomes more likely, so calls tend to be terminated before many crossings can be made. It is also seen that as C_h increases, hand-off activity is increased because calls are less likely to be terminated early.

In Figures 4.1 and 4.2 the effects of imperfect detection of hand-off need is considered. The missed hand-off initiation probability was taken as 1.0E-03 and 1.0E-05, respectively. In either case, blocking probability is insensitive to this parameter over the interesting range of demand. However, for forced termination probability, it is seen that in the low demand region, the missed detection probability determines the forced termination probability. This is because few calls are terminated due to lack of resources in the target cell. For high demands the lack of resources dominates the forced termination probability so the curves tend to follow those for perfect detection. Thus the missed detection probability sets a "floor" on forced termination performance. Significant improvement in P_{FT} by increasing C_h is only attainable in the high demand region. Figure 4.3 shows blocking and forced termination probabilities as a function of dwell time. The figure includes the effect of missed hand-off initiations ($\zeta = 1.0E{-}03$ was assumed). The plot is for $\Lambda(1) = 2.75E{-}04$ calls/sec., - the middle range of Figure 4.1. The corresponding "perfect detection" figure is Figure 1.3. Performance degradation from that case is seen by comparing the two.

Parameter Sets Used

PARAMETER SET A

$C = 15$, $C_h = 0,2,4$, $G = 2$, $N(1) = 2$, $N(2) = 2$, $v(1,0) = 150$,

$\alpha_n(2) = \Lambda_n(2)/\Lambda_n(1) = 1.0$, $\overline{T}(1) = \overline{T}(2) = 100$ sec.,

$\overline{T}_D(1,1) = \overline{T}_D(1,2) = 500$ sec., $\overline{T}_D(1) = 1000$ sec.,

$\overline{T}_D(2,1) = \overline{T}_D(2,2) = 100$ sec., $\overline{T}_D(2) = 200$ sec.

PARAMETER SET B

$C = 15$, $C_h = 0,2,4$, $G = 2$, $N(1) = 2$, $N(2) = 2$, $v(1,0) = 150$,

$\Lambda(1) = 2.75E{-}04$, $\alpha_n(2) = \Lambda_n(2)/\Lambda_n(1) = 1.0$, $\overline{T}(1) = \overline{T}(2) = 100$ sec.,

$\overline{T}_D(1,1) = \overline{T}_D(1,2)$, $\overline{T}_D(2,1) = \overline{T}_D(2,2)$,

$\overline{T}_D(2) \, / \, \overline{T}_D(1) = 0.2$.

PARAMETER SET C

$C = 15$, $C_h = 0,2,4$, $G = 1$, $N(1) = 1,2,3,4$, $v(1,0) = 300$,

$\Lambda(1) = 2.75E{-}04$, $\overline{T}(1) = 100$ sec., $\overline{T}_D(1,i) = \overline{T}_D(1,j)$

PARAMETER SET D
Same as *parameter set A*, except that $C = 25$.

PARAMETER SET E
Same as *parameter set B*, except that $C = 25$.

TABLE 3: PARAMETER SETS USED IN FIGURES

Figures:	0.1	1.1	1.2	1.3	1.4	1.5	2.1	2.2	2.3	2.4	2.5	3.1	3.2
Sets:	---	A	A	B	B	B	D	D	E	E	E	C	C

Figures:	4.1	4.2	4.3
Sets:	A, $\zeta = 1.0E{-}03$	A, $\zeta = 1.0E{-}05$	B, $\zeta = 1.0E{-}03$

192

REFERENCES

[1]   D.C. Cox and D.O. Reudink, "Layout and control of high-capacity systems," Chap. 7 in *Microwave Mobile Communications*, ed. by W.C. Jakes, Jr., Wiley:New York, 1974, pp. 545-622.

[2]   D. Hong and S.S. Rappaport, "Traffic model and performance analysis for cellular mobile radiotelephone systems with prioritized and non-prioritized hand-off procedures," IEEE Trans. Vehic. Technol., vol. VT-35, pp. 77-92, Aug. 1986.

[3]   S.A. El-Dolil, W-C. Wong, and R. Steele, "Teletraffic performance of highway microcells with overlay macrocell," IEEE J. Select. Areas Commun., vol. 7, no. 1, pp. 71-78, Jan. 1989.

[4]   D. Hong and S.S. Rappaport, "Priority oriented channel access for cellular systems serving vehicular and portable radio telephones," IEE (British) Proc., Part I, Commun., Speech and Vision, vol. 136, pt. I, no. 5, pp. 339-346, Oct. 1989.

[5]   R. Guerin, "Queueing-blocking system with two arrival streams and guard channels," IEEE Trans. Commun., vol. COM-36, pp. 153-163, Feb. 1988.

[6]   S.S. Rappaport, "The multiple-call hand-off problem in high-capacity cellular communications systems," IEEE Trans. Vehic. Technol., vol. 40, no. 3, pp. 546-557, Aug. 1991 .

[7]   S. Rappaport, "Models for call hand-off schemes in cellular communication networks," in *Third Generation Wireless Information Networks*, ed. by S. Nanda and D.J. Goodman, Kluwer Academic Publishers: Boston, 1992, pp. 163-185.

[8]   S.S. Rappaport, "Modeling the hand-off problem in personal communications networks," Proc. IEEE Vehicular Technology Conference, VTC '91, St. Louis, May 1991, pp. 517-523.

[9]   R.B. Cooper, *Introduction to Queueing Theory, 2nd ed.,* Elsevier North Holland:New York, 1981.

[10]   R. Steele, "Deploying personal communication networks," IEEE Communications Magazine, September 1990, pp. 12 ff.

[11]   D.C. Cox, "Personal communications - a viewpoint," IEEE Communications Magazine, November 1990, pp. 8 ff.

[12]   A.J. Viterbi, "Wireless digital communication: a view based on three lessons learned," IEEE Communications Magazine, September 1991, pp. 33 ff.

[13]   S.S. Rappaport, "Blocking, hand-off and traffic performance for cellular communication systems with mixed platforms," CEAS Technical Report no. 610, College of Engineering and Applied Sciences, State University of New York, Stony Brook, New York 11794-2350, November 27, 1991, Appendices A and B.

[14]   A. Murase, I.C. Symington, and E. Green, "Handover criterion for macro and microcellular systems," Proc. IEEE Vehicular Technology Conference, VTC '91, St. Louis, May 1991, pp. 524-530.

[15]   S.T.S. Chia, "The control of handover initiation in microcells," Proc. IEEE Vehicular Technology Conference, VTC '91, St. Louis, May 1991, pp. 531-536.

[16]   M. Gudmundson, "Analysis of handover algorithm," Proc. IEEE Vehicular Technology Conference, VTC '91, St. Louis, May 1991, pp. 537-542.

[17]   O. Grimlund and B. Gudmundson, "Handoff strategies in microcellular systems," Proc. IEEE Vehicular Technology Conference, VTC '91, St. Louis, May 1991, pp. 505-510.

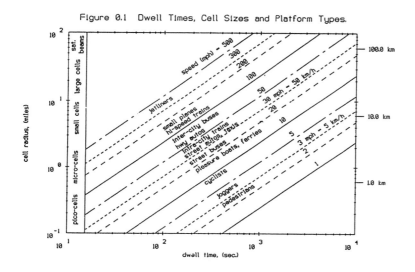

Figure 0.1   Dwell Times, Cell Sizes and Platform Types.

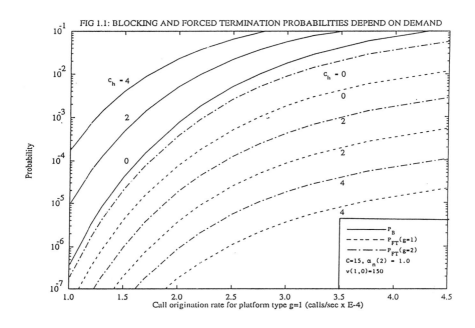

FIG 1.1: BLOCKING AND FORCED TERMINATION PROBABILITIES DEPEND ON DEMAND

194

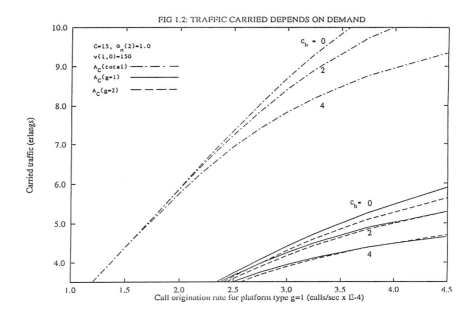

FIG 1.2: TRAFFIC CARRIED DEPENDS ON DEMAND

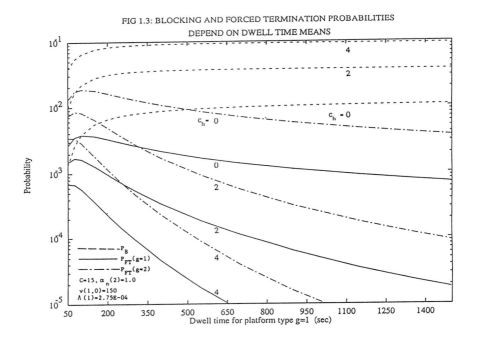

FIG 1.3: BLOCKING AND FORCED TERMINATION PROBABILITIES
DEPEND ON DWELL TIME MEANS

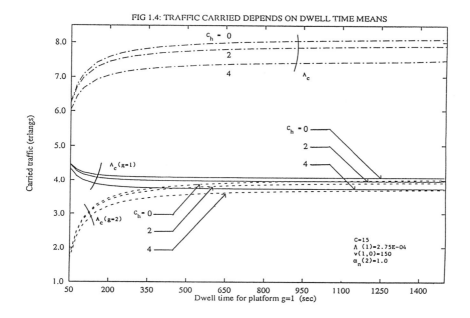

FIG 1.4: TRAFFIC CARRIED DEPENDS ON DWELL TIME MEANS

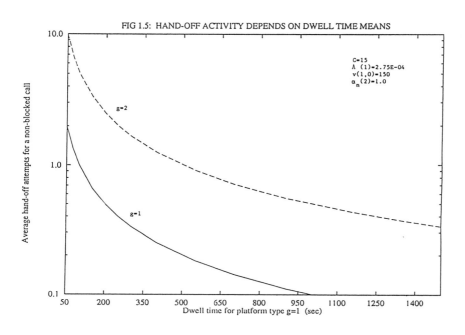

FIG 1.5: HAND-OFF ACTIVITY DEPENDS ON DWELL TIME MEANS

196

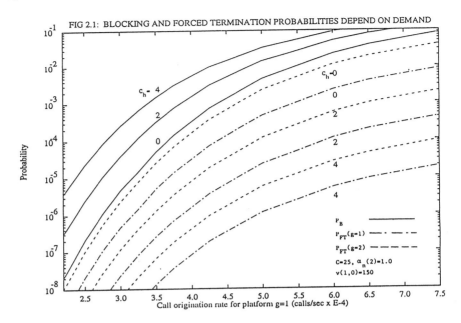

FIG 2.1: BLOCKING AND FORCED TERMINATION PROBABILITIES DEPEND ON DEMAND

Probability

$c_h = 4$

2

0

$c_h = 0$

0

2

2

4

4

P_B ——————
$P_{FT}(g=1)$ —·—·—·
$P_{FT}(g=2)$ ————
$C=25, \alpha_n(2)=1.0$
$v(1,0)=150$

Call origination rate for platform g=1 (calls/sec x E-4)

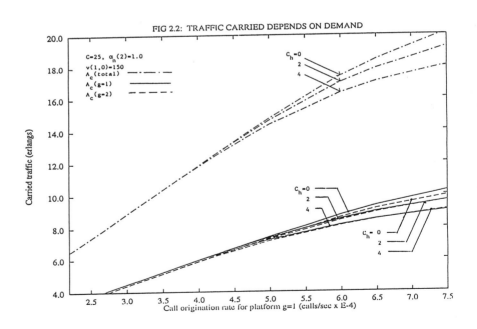

FIG 2.2: TRAFFIC CARRIED DEPENDS ON DEMAND

Carried traffic (erlangs)

$C=25, \alpha_n(2)=1.0$
$v(1,0)=150$
$A_c(total)$ —·—·—
$A_c(g=1)$ ——————
$A_c(g=2)$ ————

$c_h = 0$
2
4

$c_h = 0$
2
4

$c_h = 0$
2
4

Call origination rate for platform g=1 (calls/sec x E-4)

197

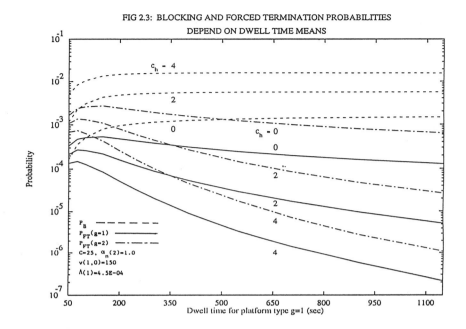

FIG 2.3: BLOCKING AND FORCED TERMINATION PROBABILITIES
DEPEND ON DWELL TIME MEANS

$c_h = 4$

2

0

$c_h = 0$

0

2

2

4

4

Probability

P_B — — —
$P_{FT}(g=1)$ ————
$P_{FT}(g=2)$ —·—·—
$C=25$, $\alpha_n(2)=1.0$
$v(1,0)=150$
$\Lambda(1)=4.5E-04$

Dwell time for platform type g=1 (sec)

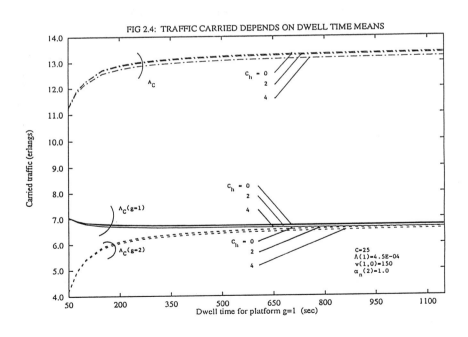

FIG 2.4: TRAFFIC CARRIED DEPENDS ON DWELL TIME MEANS

A_C

$c_h = 0$
2
4

$c_h = 0$
2
4

$A_C(g=1)$

$A_C(g=2)$

$c_h = 0$
2
4

$C=25$
$\Lambda(1)=4.5E-04$
$v(1,0)=150$
$\alpha_n(2)=1.0$

Carried traffic (erlangs)

Dwell time for platform g=1 (sec)

198

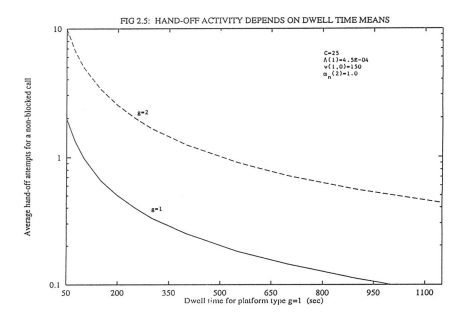

FIG 2.5: HAND-OFF ACTIVITY DEPENDS ON DWELL TIME MEANS

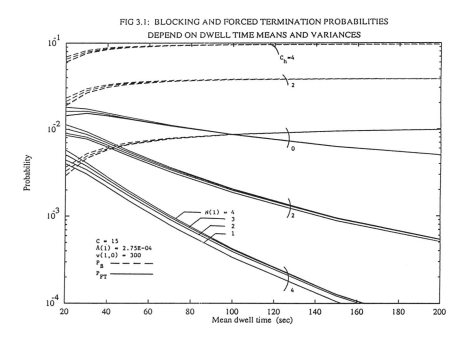

FIG 3.1: BLOCKING AND FORCED TERMINATION PROBABILITIES
DEPEND ON DWELL TIME MEANS AND VARIANCES

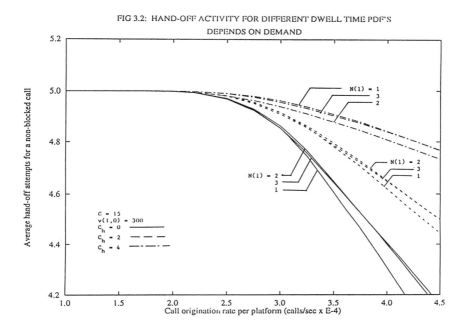

FIG 3.2: HAND-OFF ACTIVITY FOR DIFFERENT DWELL TIME PDF'S
DEPENDS ON DEMAND

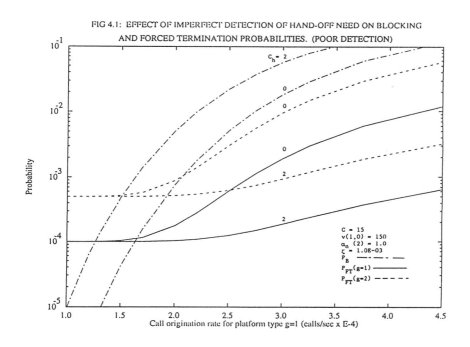

FIG 4.1: EFFECT OF IMPERFECT DETECTION OF HAND-OFF NEED ON BLOCKING
AND FORCED TERMINATION PROBABILITIES. (POOR DETECTION)

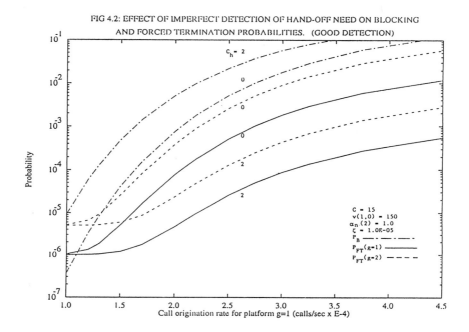

FIG 4.2: EFFECT OF IMPERFECT DETECTION OF HAND-OFF NEED ON BLOCKING AND FORCED TERMINATION PROBABILITIES. (GOOD DETECTION)

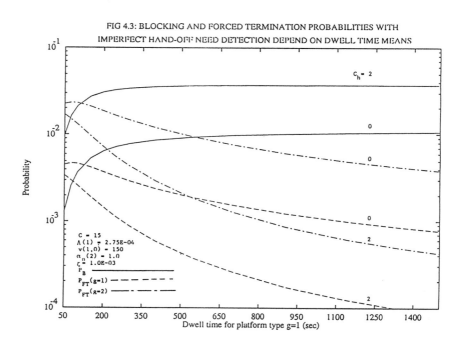

FIG 4.3: BLOCKING AND FORCED TERMINATION PROBABILITIES WITH IMPERFECT HAND-OFF NEED DETECTION DEPEND ON DWELL TIME MEANS

Traffic Analysis of Co-Channel Interference in TDMA Personal Communications Systems: Theoretical Framework

E.H. Lipper and M.P. Rumsewicz
Bellcore
331 Newman Springs Rd
Red Bank, NJ, 07701

Abstract

This paper presents an analytical modeling framework for studying the impact of co-channel interference on traffic performance in TDMA digital portable radio systems. The theoretical approach is based on a two-dimensional finite state Markov chain description of a base station which tracks both the number of active calls and the number of time slots unusable because of excessive co-channel interference. By suitably choosing certain key model parameters a variety of different radio environments could be studied. Typical of the traffic performance measures that can be obtained with this methodology are new call blocking probabilities, call dropping rates, and the distribution of unusable time slots resulting from co-channel interference (a measure of reduced service capacity). Example numerical calculations illustrating this approach are presented.

1. Introduction

Time-division multiple-access (TDMA) is one of a number digital radio technologies being actively investigated today [9] for wireless access to personal communications systems (PCS). One of the important issues affecting the viability of any radio-access technology will be call carrying capacity, and in the case of TDMA, it has been recognized that co-channel interference may be a limiting factor affecting the capacity of TDMA-based systems [7,10]. Co-channel interference refers to transmission impairments due to calls using the same time slot and frequency channel in areas served by nearby base-stations. Suitable models accounting for its impact on traffic performance will be needed both for design assessment as well as for eventual traffic engineering purposes. This paper presents an analytical modeling approach designed to study the impact of co-channel interference on traffic performance in a TDMA system. The success of PCS, of course, will depend critically on services to be provided by the "intelligent network", i.e. the public switched voice network and the common channel signaling network.

Hence models that can characterize traffic performance in the "wireless portion" of a PCS environment will also be important in assessing the overall impact of PCS traffic on the public network. More specifically, several of the performance measures that can be computed by our approach can be used to characterize traffic to end-offices, the signaling network, etc, and hence determine the overall traffic impact of PCS.

Before describing the model, we briefly discuss the operation of a basic TDMA-based system. Portable units communicate with a base station on a specific frequency channel which is time-divided into a number of fixed rate slots. The number of time slots is an upper bound on the number of calls that can be simultaneously active on a single base station frequency. At call setup, a portable unit requests a time slot from the base station. If no idle time slots are available when the request is made, the call would be blocked.[1] Since the area served by a single base station is likely to be very small compared to those in vehicular mobile systems, [1-4,9] the same frequency may be used by base stations that are very closely spaced. The spatial arrangement of the base stations in the network, and the total number of frequencies available would determine the number (and location) of nearby base stations significantly affecting a given base station through co-channel interference. (This is discussed further in Section 3). In addition to affecting transmission quality, co-channel interference could affect traffic performance in two important ways. First, time slots experiencing unacceptable interference levels might not be assigned to new calls, even if they were the only remaining idle slots [7,11]. This means that calls could be blocked, even though idle slots exist. Second, the interference level will vary over time, depending on the process of call originations and completions on the same frequency channel-time slot pair at nearby base stations, i.e. on the number of "co-channel interferers". If the interference becomes unacceptable on a time slot, it may become necessary to transfer a call in progress either to another time slot on the base station's frequency, or to

---

1. The particular time slot assigned to the portable for a call might be selected to meet certain transmission quality criteria, [11]. We are basically modeling the "uplink" portion of call access from the portable to the base station. Depending on the technology to be implemented, the portable units may attempt to obtain time slots from nearby base stations, if the initial base station's time slots are all unavailable. See [12,13]. In this case the call would not be blocked unless no time slots were available at the nearby base station as well. This alternative is considered in [17].

a different base station entirely. This would be done in order to maintain the transmission quality of the call. Of course, call transfers may also result from user motion, as is common in the vehicular mobile environment, [9]; but co-channel interference or other impairments may be the dominant cause of transfers in PCS since users will either be stationary or moving very slowly. Regardless of whether the transfer attempt is within a base station, or to another base station, a new time slot must be found that is both *idle*, in the traffic sense (assuming call preemption is not permitted), and *"usable"* from an interference standpoint. Of course, when a call is transferred in this way, the co-channel interference levels of both "old" and "new" time slots are affected. Failure to find an available time slot for a transfer might cause the call to be terminated prematurely. It is important therefore to characterize transfer traffic so that transfer blocking probabilities can be determined.

Most attempts to model these phenomena have relied on detailed simulations of the radio propagation environment in which relatively simple traffic models have been embedded [7, 11-13]. In a recent paper, [14], analytical modeling of the radio propagation environment combined with independent use of the Erlang-B methodology for computing blocking probabilities is proposed as a means for studying co-channel interference effects. The analytical methodology proposed in this paper is a more sophisticated characterization of the randomly varying state of a base station, accounting for the effect that call traffic at nearby base stations has on co-channel interference in the base station under study. We are thus able to compute various performance measures, including blocking probability, number of unusable time slots due to co-channel interference, the rate at which calls are dropped, etc.

The remainder of this paper is organized as follows. The next section introduces the theoretical framework which, depending on the choice of certain key parameters, allows for a variety of scenarios to be studied. After illustrating the methodology with several simple examples, Section 3 discusses a parameter selection methodology for a more realistic base station arrangement affecting the base station under study. Section 4 provides some numerical examples illustrating the type of results that can be obtained using this approach, while Section 5 summarizes the contributions of this paper.

2. Mathematical Model

The theoretical framework is based on a 2-dimensional, finite state Markov Chain description of a base station. The model keeps track of the number of busy slots, as well as the number of slots in the "unusable" state. A slot is considered *"available"* if it is both idle and usable, and new call attempts are lost if no available slots exist. A call in progress is transferred to an available slot, if one exists, when its current slot moves to the unusable state. This transfer corresponds to an intra base station transfer. If no available slot exists, such a call is lost from the system. In practice an inter-base station transfer for this call would be attempted. Hence this methodology can be used to calculate not only the new call blocking probability, but also to characterize the inter-base station transfer traffic generated by the traffic offered to the base station under study.

Consider a TDMA-based base station, in which each call requires exactly one time slot for transmission. Let S be the number of slots in the TDMA frame and hence the maximum number of users served by a base station that can simultaneously be involved in calls. We introduce a random variable, K, to keep count of the number of unusable slots due to interference. Clearly $S - K$ is the maximum number of calls that can be in progress in the base station when K channels are unusable. The random variable K can take values from the set $\{0,...,S\}$ and will be assumed to have a skip free property (that is, can change by at most one when it changes). While this restriction is not necessary for the analysis, it provides a more reasonable modeling base because the level of interference for a specific slot will change primarily because of new calls starting or old calls finishing on the same frequency channel and time slot in other base stations. As these occur one at a time it is reasonable to assume the K will change similarly. Intra-base station transfers are assumed to be attempted when a slot having a call in progress becomes unusable.

We define the state of a base station by the pair (N, K), where K is as described earlier and N is the number of calls in progress in the base station, with $N \leq S - K$. We now describe the manner in which state changes can occur from state (n,k). For $n < S - k$ new calls occur at rate λ_n, causing the state to change from (n,k) to $(n+1,k)$. If $n = S - k$ no arrivals are accepted to the system because there are no available slots for the call. Calls in progress can be terminated for two reasons: (1) the call completes or (2) the time slot becomes unusable and no other slots are available. In the first case, the state moves from (n,k) to $(n-1,k)$ at rate μ_n. In the second case, the state moves from $(n,S-n)$ to $(n-1,S-n+1)$ at rate $\gamma_{n,S-n}$. Finally, changes in the interference environment, K, cause the state to change from (n,k) to $(n,k+1)$ at rate $\gamma_{n,k}$ and to $(n,k-1)$ at rate $\alpha_{n,k}$. Figure 1 shows the state transition diagram summarizing the possible state changes.

Figure 1. State transition diagram (for $S = 6$)

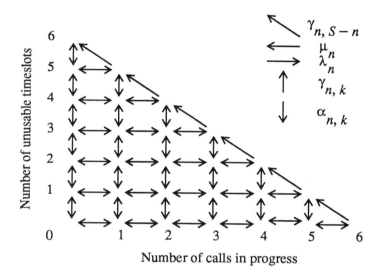

Number of unusable timeslots

Number of calls in progress

The infinitesimal generator of this system has a block tridiagonal structure, [15], [16], and the global balance equations can be written in the form:

$$\mathbf{p}_0 A_1^{(0)} + \mathbf{p}_1 A_2^{(1)} = 0 \qquad (2.1)$$

$$\mathbf{p}_{n-1} A_0^{(n-1)} + \mathbf{p}_n A_1^{(n)} + \mathbf{p}_{n+1} A_2^{(n+1)} = 0, \quad n = 1,...,S-1$$

$$\mathbf{p}_{S-1} A_0^{(S-1)} + \mathbf{p}_S A_1^{(S)} = 0$$

where $\mathbf{p}_n = (p_{n,0}, p_{n,1}, \cdots, p_{n,S-n})$ and $p_{n,k} = Pr\{N = n, K = k\}$.
$A_0^{(n)}, (n = 0,...,S-1)$ is an $(S-n+1) \times (S-n)$ matrix with

$$A_0^{(n)}(i,j) = \begin{cases} \lambda_n & \text{if } i = j, i = 0,...S-n-1 \\ 0 & \text{otherwise} \end{cases} \qquad (2.2)$$

$A_2^{(n)}, (n = 1,...,S)$ is an $(S-n+1) \times (S-n+2)$ matrix with

$$A_2^{(n)}(i,j) = \begin{cases} \mu_n & \text{if } i = j,\, i = 0,...,S-n \\ \gamma_{n,S-n} & \text{if } i = S-n,\, j = S-n+1 \\ 0 & \text{otherwise} \end{cases} \quad (2.3)$$

$A_1^{(n)},(n = 0,...,S)$ is an $(S-n+1) \times (S-n+1)$ matrix with, for $i \neq j$,

$$A_1^{(n)}(i,j) = \begin{cases} \gamma_{n,i} & \text{if } j = i+1,\, i \neq S-n \\ \alpha_{n,i} & \text{if } j = i-1,\, i \neq 0 \\ 0 & \text{otherwise} \end{cases} \quad (2.4)$$

and for $i = j$,

$$A_1^{(n)}(i,j) = -[\lambda_n \, I_{\{i \neq S-n\}} + \mu_n I_{\{n \neq 0\}} + \alpha_{n,i} I_{\{i \neq 0\}} + \gamma_{n,i} I_{\{i \neq S\}}], \quad (2.5)$$

where I is the indicator function.

As in [16] the solution can be determined numerically by the following scheme:

Define $C^{(0)} = A_1^{(0)}$.

Compute

$$C^{(n)} = A_1^{(n)} + A_2^{(n)}(-(C^{(n-1)})^{-1})A_0^{(n-1)}, \quad \text{for } n = 1,...,S. \quad (2.6)$$

Solve $\mathbf{p}_S \, C_S = 0$.

Compute

$$\mathbf{p}_n = \mathbf{p}_{n+1}A_1^{(n+1)}(-(C^{(n)})^{-1}), \quad \text{for } n = S-1,...,0. \quad (2.7)$$

Renormalize so that

$$\sum_{n=0}^{S}\sum_{k=0}^{S-n} p_{nk} = 1. \quad (2.8)$$

A number of performance measures can now be expressed in terms of the equilibrium distribution just determined. The call blocking probability, p_B, is given by

$$p_B = \frac{\sum\limits_{n=0}^{S} \lambda_n p_{n,S-n}}{\sum\limits_{n=0}^{S} \sum\limits_{k=0}^{S-n} \lambda_n p_{n,k}}. \tag{2.9}$$

The rate at which calls in progress are lost due to the unavailability of any usable slots when their current slot becomes unusable, is given by,

$$R_D = \sum_{n=1}^{S} \gamma_{n,S-n} p_{n,S-n}. \tag{2.10}$$

The mean number of unusable slots, a measure of "lost service capacity", is given by

$$\bar{N}_u = \sum_{n=0}^{S} \sum_{k=1}^{S-n} k p_{n,k} \tag{2.11}$$

The rate at which calls are transferred from one slot to another, because of interference, is a measure of the amount of work the system must perform to maintain customer service standards. This rate, β, is given by

$$\beta = \sum_{n=1}^{S} \sum_{k=0}^{S-n-1} p_{n,k} \gamma_{n,k} \frac{n}{S-k}. \tag{2.12}$$

We now provide some examples to demonstrate the selection of $\gamma_{n,k}$, $\alpha_{n,k}$ in simple cases. In Section 3 we describe their selection for more realistic scenarios.

The M/G/S-Loss System (Erlang Loss System)

A commonly used model for determining the blocking performance of TDMA-based base station [9] is the M/G/S-loss system (or Erlang Loss System). This model does not account for co-channel interference. New calls arrive in a Poisson stream (rate λ). If a slot is available the call is accepted, and a channel assigned and held for a generally distributed length of time (with mean μ^{-1}), otherwise the call is lost. The solution for this system is well known, but nonetheless it is a special case of our construction if parameters are chosen appropriately, specifically with $\lambda_n = \lambda$, $\mu_n = n\mu$, $\alpha_{n,k} = \gamma_{n,k} = 0$. Clearly with these parameter settings, K will not change, so for completeness we define $K = 0$ always. Note that all of the matrices requiring inversion in our analysis remain non-singular with this choice of parameters.

Finite source model with no co-channel interference

This model is similar to that described above, but differs in that a source generating calls is associated with each subscriber using a base station. This is the quasirandom input loss model which has also been applied in some studies of TDMA systems, [11]. The source generates calls at rate δ if the user is idle and stops generating calls when the user is busy. If M subscribers are using the base station, the parameters for our model are chosen as in the Erlang Loss System but with $\lambda_n = (M-n)\delta$, where n is the number of calls in progress. Again, K will not change so is defined to be 0 always.

Exactly two base stations interacting through co-channel interference

Suppose that only two base stations are using the same frequency (in other words, no other base stations act as a source of co-channel interference). Suppose also that the interference is sufficient that if time slot i at one base station is being used, slot i at the other base station is unusable. Suppose also that there are M_i subscribers using base station i ($i = 1,2$) each generating calls at rate δ_i when they are idle, and the holding time of such calls has mean v_i^{-1}. Using the parameters defined in the previous section, we have, for the base station under study,

$$\lambda_n = (M_1-n)\delta_1,$$

$$\mu_n = n v_1,$$

$$\alpha_{n,k} = k v_2,$$

$$\gamma_{n,k} = \begin{cases} (M_2 - k)\delta_2 & \text{if } M_2 - k < S \\ 0 & \text{otherwise} \end{cases}$$

Note that in this example the variable k essentially keeps count of the number of time slots in use at the other base station.

3. Determination of environment parameters

The previous section provided examples on selection of the environment parameters in some simple cases so as to illustrate the methodology. In this section we focus more closely on their selection in more realistic scenarios; in particular, when the base station under study experiences interference from a number of surrounding base stations using the same frequency.

We shall assume that interference is only caused by base stations comprising what is often referred to as the "first interference tier," [9]. We denote the number of interfering surrounding base stations by T. Figure 2 shows a 3-frequency reuse pattern, in which case $T = 6$, i.e., significant co-channel interference is only caused by 6 base stations.

Figure 2. Illustration of 3 frequency reuse pattern and connection to local network.

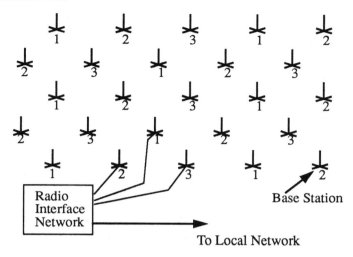

Denote by $q(j)$ the probability that a given time slot is in use in exactly j first tier base stations, i.e. that there are exactly j co-channel interferers. The quantities $q(j)$ can be determined in the following fashion. Label the first tier base stations $1,...,T$ and define $\Phi = \{1,...,T\}$. Let $A_c^{(i)}$, $i \in \Phi$, be the carried load for first tier base station i, and define $a_c^{(i)} = A_c^{(i)}/S$.

Let $\Phi(j)$ be the set of all j member subsets of Φ. Then, assuming independence between the base stations and that the first tier base stations are not subject to interference,

$$q(j) = \sum_{\chi \in \Phi(j)} \left[\left[\prod_{i \in \chi} a_c^{(i)} \right] \left[\prod_{i \in \Phi - \chi} (1 - a_c^{(i)}) \right] \right]. \qquad (3.1)$$

The assumption that the first interference tier base stations are not themselves subject to interference is relaxed in [17].

Then, if we were to examine the process underlying slot i of base station j, we would see the following: firstly, periods of conversation with mean τ_j, the mean conversation length in base station j, followed by idle periods with mean length κ_j. The τ_j's are known *a priori*, and the κ's can be determined from:

$$\kappa_j = \tau_j \, \frac{1-a_c^{(j)}}{a_c^{(j)}}. \tag{3.2}$$

Hence the mean rate at which slot i moves from the idle state to the conversation state is $1/\kappa_j$.

Consequently, the rate at which the number of base stations, with slot i in use, increases from n to $n+1$, denoted by $r_{n,n+1}$, is

$$r_{n,n+1} = \frac{1}{q(n)} \sum_{\chi \in \Phi(n)} \left[\left[\prod_{i \in \chi} a_c^{(i)} \right] \left[\prod_{i \in \Phi-\chi} (1-a_c^{(i)}) \right] \sum_{j \in \Phi-\chi} \frac{1}{\kappa_j} \right]. \tag{3.3}$$

In a similar fashion, the rate at which the number of base stations, with slot i in use, decreases from n to $n-1$, denoted $r_{n,n-1}$, is

$$r_{n,n-1} = \frac{1}{q(n)} \sum_{\chi \in \Phi(n)} \left[\left[\prod_{i \in \chi} a_c^{(i)} \right] \left[\prod_{i \in \Phi-\chi} (1-a_c^{(i)}) \right] \sum_{j \in \chi} \frac{1}{\tau_j} \right]. \tag{3.4}$$

Now, let $s_{j,k}^{v,w}$, $j=0,...,T$, v, $w=0,1$ be the conditional probability that the slot goes from state v when in use in j first tier base stations, to state w when in use at k first tier base stations, where v, $w=0$ means the slot is unusable and v, $w=1$ means the slot is usable. Discussion on the choice of these probabilities is given later in the paper.

For the interference environment process governing slot i, we create an intermediate process X. The state space of X consists of the pairs (j,v), where $j=0,...,T$ and $v=0,1$, where v is defined as above. In this way we keep track of the number of first tier base stations using slot i, as well as whether or not the slot is usable.

Let $p'(j,v)$ denote the probability that the process X is in state (j,v). The state transition diagram is given in Figure 3 and the global balance equations governing the process X^2 are:

---

2. Note: Other choices for the process X are, of course, possible. If the process X is modified by allowing transitions between states $(j,0)$ and $(j,1)$, movement of subscribers within base station coverage areas would be immediately accounted for. Suitable models for customer movement and its effect on the interference environment would then be needed. However, each physical environment to which this is applied would lead to different choices of the appropriate rates. As this is still an open question undergoing research, we have chosen a simpler formulation with a coarser state description that still captures the main features of the interference environment process.

$$p'(0,1)\, r_{0,1} = p'(1,1)\, r_{1,0}\, s^{1,1}_{1,0} + p'(1,0)\, r_{1,0}\, s^{0,1}_{1,0} \qquad (3.5)$$

$$p'(j,1)(r_{j,j+1} + r_{j,j-1}) = p'(j-1,1)\, r_{j-1,j}\, s^{1,1}_{j-1,j} \qquad (3.6)$$
$$+ p'(j-1,0)\, r_{j-1,j}\, s^{0,1}_{j-1,j} + p'(j+1,1)\, r_{j+1,j}\, s^{1,1}_{j+1,j}$$
$$+ p'(j+1,0)\, r_{j+1,j}\, s^{0,1}_{j+1,j}$$
$$\text{for } 1 \le j \le T-1$$

$$p'(T,1)\, r_{T,T-1} = p'(T-1,1)\, r_{T-1,T}\, s^{1,1}_{T-1,T} \qquad (3.7)$$
$$+ p'(T-1,0)\, r_{T-1,T}\, s^{0,1}_{T-1,T}$$

$$p'(0,0)\, r_{0,1} = p'(1,1)\, r_{1,0}\, s^{1,0}_{1,0} + p'(1,0)\, r_{1,0}\, s^{0,0}_{1,0} \qquad (3.8)$$

$$p'(j,0)(r_{j,j+1} + r_{j,j-1}) = p'(j-1,1)\, r_{j-1,j}\, s^{1,0}_{j-1,j} \qquad (3.9)$$
$$+ p'(j-1,0)\, r_{j-1,j}\, s^{0,0}_{j-1,j} + p'(j+1,0)\, r_{j+1,j}\, s^{0,0}_{j+1,j}$$
$$+ p'(j+1,1)\, r_{j+1,j}\, s^{1,0}_{j+1,j}$$
$$\text{for } 1 \le j \le T-1$$

$$p'(T,0)\, r_{T,T-1} = p'(T-1,1)\, r_{T-1,T}\, s^{1,0}_{T-1,T} \qquad (3.10)$$
$$+ p'(T-1,0)\, r_{T-1,T}\, s^{0,0}_{T-1,T}$$

Figure 3. State transition diagram for the X process (with $T = 6$).

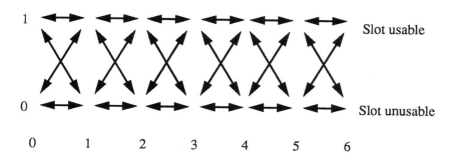

Number of base stations with a given time slot in use

Solving these equations numerically, we can then obtain the average rate at which slot i moves from usable to unusable, and *vice-versa*, namely:

$$u(1,0) = \frac{\sum_{j=0}^{T-1} \left[p'(j,1) \, (r_{j,j+1} \, s_{j,j+1}^{1,0}) + p'(j+1,1)(r_{j+1,j} s_{j+1,j}^{1,0}) \right]}{p'(\bullet,1)} \qquad (3.11)$$

$$u(0,1) = \frac{\sum_{j=1}^{T} \left[p'(j,0) \, (r_{j,j-1} \, s_{j,j-1}^{0,1}) + p'(j-1,0)(r_{j-1,j} s_{j-1,j}^{0,1}) \right]}{p'(\bullet,0)} \qquad (3.12)$$

where $p'(\bullet,i)$ is the marginal probability that the slot is usable ($i=1$) or unusable ($i=0$). Now we can write down the γ's and α's defined earlier. In particular,

$$\gamma_{n,k} = (S - k)u(1,0) \qquad (3.13)$$
$$\alpha_{n,k} = ku(0,1) \qquad (3.14)$$

These follow since the rate at which new slots become unusable is directly proportional to the number of slots which are usable ($S - k$). Similarly, the rate at which slots become usable is directly proportional to the number of unusable slots (k), as each slot is assumed independent of all others.

Note that we have made the interference environment parameters independent of the number of calls in progress in the base station in question. From a physical point of view this is appropriate since the interference environment is determined by the active calls in first tier base stations, not the calls in progress in the base station of interest.

Selection of transition probabilities for process X

A decision must also be made on the nature of the probabilities $s_{j,j'}^{x,x'}$. Consider time slot i of the base station under study. The interference level experienced by this time slot is due to its use in the base stations of the first tier. If this time slot is in use in j of the first tier base stations, let b_j denote the probability that the level of interference experienced in the time slot is beyond the acceptable level. In other words, b_j is the probability that the time slot is unusable given that there are currently j co-channel interferers.

These probabilities depend on detailed characteristics of the radio environment which will depend on base station coverage area, terrain, building characteristics, power and other technological issues. By "fitting" either measurement data, analytical results (as in [14]), or simulation results (as in [11]), as these results become available from ongoing research, our theoretical framework would be able to accommodatea variety of scenarios.

For example, one potential choice for the state transition probabilities of the X process would be to set

$$s_{j,j\pm1}^{*,0} = b_{j\pm1}, \quad \text{for } 0 \le j \le T \tag{3.15}$$

This ensures that, given a certain number of co-channel interferers, the chance of a slot being unusable is appropriately accounted for. In this case, we have essentially introduced a biased coin toss whenever the number of base stations using a particular slot changes. Physically, one can consider this as taking into account the fact that subscribers may move around (and therefore change the interference environment), but only accounting for it mathematically at certain jump times.

Other choices of $s_{j,x}^{y,w}$. can be made depending on the characteristics that one wishes the process X to exhibit. As another example, if it is required that a slot, usable when in use at j base stations, is still usable when j becomes $j-1$, $s_{j,j-1}^{1,0}$ must be set to 0. Similarly, if it is required that a slot, unusable when in use at j base stations, is still unusable when j becomes $j+1$, $s_{j,j+1}^{0,1}$ must be set to 0. The other values for the transition probabilities can then be related to certain conditional probabilities on the b_j's.

4. Numerical Experiments

In this section we provide some examples to illustrate the potential impact of co-channel interference on the number of time slots that are unusable and the subsequent effect on customer perceived grade of service. In the following discussion we will assume that there six base stations in the first interference tier (that is, $T = 6$), there are ten time slots (that is, $S = 10$), and that the mean interarrival time of new calls to the base station under study is 33.3 seconds with mean holding time of 150 seconds, thereby corresponding to a 1% blocking criteria in the absence of interference (as given by the Erlang B formula). The X process used will be that with transition probabilities described by Equation 3.15.

Figures 4, 5 and 6 each show three curves, one corresponding to an Erlang B approximation, while the other two correspond to different choices for the b_j's. Interference pattern 1 has $\mathbf{b} = (b_0, b_1, \cdots, b_6) = (0, 0.01, 0.02, 0.04, 0.08, 0.16, 0.32)$, while Interference pattern 2 has $\mathbf{b} = (0, 0.02, 0.04, 0.08, 0.16, 0.32, 0.64)$. The horizontal axis is the carried load in each of the first tier interference cells (that is, the $a_c^{(i)}$'s defined in earlier) while the vertical axis shows the performance measures (1): percent utilization of time slots (Figure 4), (2): new call blocking probability (Figure 5), and (3): percent time slots unusable due to interference (Figure 6).

As can be seen from these figures, the performance (in terms of customer grade of service) can be greatly affected by the choice of the interference parameters. Moreover, the Erlang B approximation can break down even for offered loads of the order of 50% of the system capacity (which corresponds to about 1% blocking in the case where no interference effects are accounted for).

Figure 7 shows the distribution in time for the number of unusable slots (for the two different sets of b_j's used in Figures 4, 5 and 6), when the carried load is about 0.5 erlang per slot in the first tier cells. This illustrates the fraction of time that one can expect different numbers of slots to be unusable.

Clearly these figures can be greatly affected by the choice of input parameters. We are not claiming here that the parameters selected are typical of any particular environment. However, if certain environments can be characterized in terms of the parameters b_j that we have defined, our analysis provides a method for determining the potential impact of interference on customer service as a function of traffic load.

5. Conclusion

In this paper we have presented a methodology for analyzing the impact of co-channel interference on traffic performance in a TDMA Personal Communications System. This methodology models the effects of a randomly changing interference environment experienced by a single base station due to active calls at nearby base stations using the same frequency. By incorporating this feature, the analysis more realistically accounts for changes in signal interference levels, and thus provides more reasonable estimates for traffic performance measures than simpler techniques, such as the Erlang Loss Model. A number of important performance measures, such as call blocking probability and the rate at which calls are dropped by a base station due to interference, can be evaluated using this approach. In addition, the distribution for the number of unusable time slots, a measure of "lost service capacity", can be computed.

This approach allows more efficient computation of traffic performance measures than detailed simulation, while still incorporating radio environment characteristics. The use of such simulations or field measurements, however, is intrinsic to the application of this technique,

Figure 4. Percent utilization versus interfering base station traffic

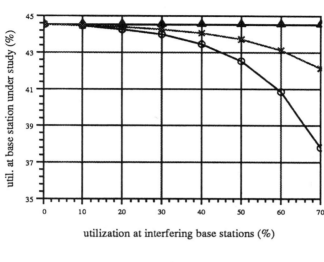

Figure 5. Percent blocking versus interfering base station traffic

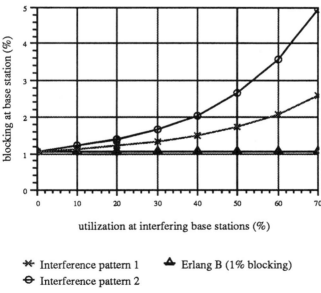

as they provide some of the necessary input parameters.

Figure 6. Percent timeslots unusable versus interfering base station traffic

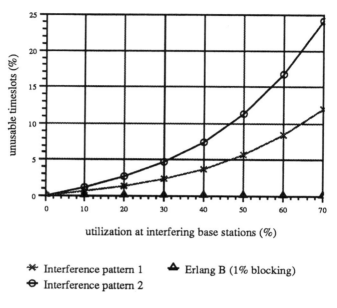

utilization at interfering base stations (%)

✳ Interference pattern 1   ▲ Erlang B (1% blocking)
⊖ Interference pattern 2

Figure 7. Equilibrium distribution for number of unusable slots

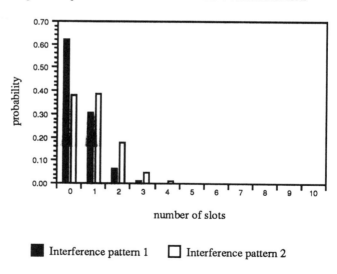

number of slots

■ Interference pattern 1   □ Interference pattern 2

In [17], we extend the technique presented here to the analysis of a network of TDMA base stations, rather than a single base station, taking into account factors such as subscriber movement, transfer traffic, and overflow of blocked calls to nearby base stations.

References

[1] D.C. Cox, "Universal digital portable radio communications," Proc. IEEE, vol. 75, pp. 436-477, Apr. 1987.

[2] -----, "Portable digital radio communications - An approach to tetherless access," IEEE Commun. Mag., pp. 30-40, July 1989.

[3] -----, "A radio system proposal for widespread low-power tetherless communications," IEEE Trans. Commun., vol. 39, pp. 324-335, Feb. 1991.

[4] "Generic Framework Criteria for Universal Digital Portable Communications," FA-TSY-001013, Bellcore, no.2, Dec. 1990.

[5] A.J. Motley, "Advanced cordless telecommunications service," IEEE J. Select. Areas Commun., pp. 774-782, June 1987.

[6] N.R. Sollenberger, J. C-I Chuang, L.F. Chang, S. Ariyavisitakul, and H.W. Arnold, "Architecture and implementation of an efficient and robust TDMA frame structure for digital portable communications," Proc. IEEE VTC'89, San Francisco, CA., April 28-May 3, 1989, pp. 169-174.

[7] R. Bernhardt, "Time-slot management in digital portable radio systems," IEEE Tran. Vehicul. Tech., vol. 40, pp. 261-272, Feb. 1991.

[8] M.R.L. Hodges, "The GSM radio interface," Br. Telecom. Technol. J., vol. 8, pp. 31-43, Jan. 1990.

[9] W. C. Y. Lee, Mobile Cellular Telecommunications Systems, McGraw-Hill Book Company, New York, 1989.

[10] D.C. Cox,"Co-channel interference considerations in frequency-reuse small- coverage area radio systems," IEEE Trans. Commun., vol. COM-30, pp. 135-142, Jan. 1982.

[11] R. Bernhardt, "User access in portable radio systems in a co-channel interference environment," IEEE J. Select. Areas Commun., vol. 7, pp. 49-58, Jan. 1989.

[12] J.C-I Chuang, "Autonomous adaptive frequency assignment for TDMA portable radio systems," 4th Nordic Sem. Digital Land Mobile Radio Commun., June 26-28, 1990, Oslo, Norway.

218

[13] J. C-I Chuang, "Operation and performance of a self-organizing frequency assignment method for TDMA portable radio," Proc. GLOBECOM'90, pp. 1548-1552, Dec. 1990.

[14] R. Prasad and A. Kegel, "Spectrum efficiency of microcellular systems," Proc. IEEE VTC '91, St. Louis, MO, May '91, pp. 357-361.

[15] M.F. Neuts, *Matrix Geometric Solutions in Stochastic Models: An Algorithmic Approach*, The Johns Hopkins University Press, Baltimore, 1981.

[16] D.P. Gaver, P.A. Jacobs and G. Latouche, "Finite birth-and-death models in randomly changing environments", *Adv. Appl. Prob.*, 16, 1984, pp. 715-731.

[17] E.H. Lipper and M.P. Rumsewicz, "Traffic Modeling of TDMA Personal Communications Networks", in preparation.

DISTRIBUTED DYNAMIC CHANNEL ALLOCATION ALGORITHMS FOR MICROCELLULAR SYSTEMS

L. J. Cimini, Jr., G. J. Foschini and C.-L. I
AT&T Bell Laboratories
Crawford Hill Laboratory
Holmdel, New Jersey 07733-0400

Abstract

Microcellular systems offer the potential for substantial increases in capacity. However, existing frequency planning and network control are impractical. A solution to the network management problems created by the use of microcells is *dynamic channel allocation* (every channel available for use in every cell) with *decentralized control* (decisions made by the mobiles or portables rather than by a central switch). By addressing elemental situations, we show that microcellular systems can self-organize, with little loss in capacity in comparison to the best globally coordinated channel selection. Moreover, this can be done by using simple channel-allocation algorithms, with the mobile or portable unit making autonomous decisions based only on local measurements.

1. Introduction

Cellular technology has greatly enhanced the capability of coping with the increased demand for mobile communication services. By dividing the service area into cells, frequency channels may be reused many times (subject to a constraint on the allowed cochannel interference) to make efficient use of the limited spectrum [1]. However, in the near future, the demand for capacity is expected to far exceed the capabilities of present cellular systems.

A promising way to increase capacity for future wireless systems is through the use of *microcells** [2-6]. That is, small cells with lower antenna heights and lower power than conventional cellular systems. Since the geographical scope of the cell is curtailed by limiting the signal

---

* A more quantitative definition of a microcell is that the cell-site antenna is at street lamp level, the emitted power is on the order of milliwatts, and the cell radius is less than 400 m. In contrast, conventional cells typically have 50-m antenna heights, emit over a watt, and have a radius over 1 km.

power and the antenna height, the channels may be reused over much shorter distances and therefore an increase in system capacity of several orders of magnitude should be possible. However, the use of small cells creates several problems in the management of the network. In particular, the number of handoffs increases and the time to make them decreases. In addition, there will be a significant increase in the amount of signaling traffic and an increase in the complexity (and, therefore, the fragility) of the control software. More importantly, the complexity of frequency planning increases as cell size decreases. The infinite variations in local terrain features, especially in an urban microcell environment, coupled with traffic variations, render theoretical frequency planning extremely difficult if not impossible.

A solution to the network management problems created by the use of microcells is *dynamic channel allocation* (DCA) with *decentralized* (or *distributed*) *control*. DCA, where every channel† is available for use in every cell, removes the need for frequency planning while adapting to local interference and traffic conditions. However, DCA is classically performed by a central switch which could be easily overloaded in a microcellular system. Distributed control, where decisions are made by the mobiles and/or bases rather than a central switch, can remove much of the burden on the switch and allows more handoffs to be accommodated faster. Keep in mind that DCA is being proposed to cope with the network planning problems imposed by microcells rather than to get additional capacity. The concept of DCA with distributed control is being emphasized by several groups throughout the world (for example, see [7-12]). In particular, two systems proposed for use in Europe in the next couple of years, DECT [11] and DCT 900 [12], implement such a scheme.*

How does a system using DCA with distributed control perform in a microcellular environment? What channel-allocation algorithm meets the performance criteria and, more importantly, how do we evaluate these algorithms? How decentralized can the control be? These issues appear highly complex and best approached by solving a series of *elemental* problems that build *incrementally* to the desired problem. That is, to build the understanding and insight necessary to design such systems, we start by looking at smaller, more-focused problems.

---

† When we speak of "channels" in this paper, we mean frequencies and/or time slots.

* In [13], a self-organizing procedure is described for developing an efficient frequency-reuse plan without requiring centralized control.

As each elemental problem is solved, the system performance must be evaluated. Relevant performance measures are *system capacity* (or spectral efficiency) and *grade of service*, which includes call blocking and dropping rates and coverage probability. Another important performance measure, number of reconfigurations per call, may be thought of as a stability measure of channel-allocation algorithms. In a system using DCA, there is the possibility that, especially when heavily loaded, an active user may experience excessive rearrangements or handoffs. By rearrangement, here we mean an intracell handoff,* which may be initiated due to a degradation in the quality of the call. We would like to limit intracell handoffs even though these handoffs need not involve the central switch. However, by the introduction of intracell handoffs, both blocking performance and capacity can be improved (for some examples, see [17-18]). Here we initiate a study of the tradeoffs between various performance measures.

As a starting point for study, we have assumed maximum decentralization, i.e., the mobile or portable unit performs the channel selection and handoffs (both intracell and intercell) and its decisions are based solely on *local* information. This will speed up the handoffs and, depending on the system architecture, should reduce much of the signaling traffic. *What is the cost of this level of decentralization?* Here we study, without the burden of detailed propagation considerations, and initially without the traffic, the performance of simple single channel-selection algorithms†. Subsequently, we will also begin to look at multichannel algorithms with traffic.

In Section 2, we describe capacity calculations for a linear array of cells. In Section 3, similar studies are described for a planar array of hexagonal cells. Future wireless systems will most surely be a mix of linear and area microcells along with more conventional macrocells. Computer simulations are used for the studies in Sections 2 and 3. In Section 4, we give a *simulated annealing* interpretation of our findings. In Section 5, we begin the investigation of multichannel algorithms as well as the inclusion of traffic considerations. Finally, in Section 6, we present conclusions and discuss areas of interest for further study.

---

* In this study, we are ignoring intercell handoffs, that is, the transfer of a call from one base station to another base station. We assume that the number of intercell handoffs can be kept to a minimum for fast moving vehicles by either using "umbrella" cells (i.e., macrocells) [14,15] or by giving the mobile unit a higher channel-allocation priority. A study of intercell handoff probability in microcells is presented in [14].

† A summary of these elemental results appears in [19].

2. Linear Microcells

2.1 Bounds on System Capacity

Consider a linear array of cells, as shown in Figure 1, with only a *single* channel available for use.

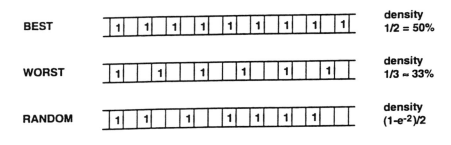

1: channel   blank: no channel

Figure 1: A Linear Array Of Cells Saturated With Channels. A structure such as this can be used to model the coverage of highways or of city streets. Best capacity: $C_{max}=50\%$. Worst capacity: $C_{min}=33\%$. Random capacity: $C_{ran} \approx 43\%$.

We do not consider the complication of traffic. Propagation and interference considerations are simply represented by the constraint that, if the channel is used in a given cell, it cannot be used in the two adjacent cells (i.e., there must be at least one buffer cell between cells using the same channel). It is evident that, if the channel assignment is *globally* coordinated, the best arrangement is for the channel to be reused in every other cell, yielding a capacity (or efficiency), $C_{max}=1/2$ (50%). In the worst case, the channel is reused in every third cell leading to a capacity, $C_{min}=1/3$ (33%). Since any completely decentralized algorithm, designed to saturate the array with channels while adhering to the reuse constraint, necessarily performs somewhere between the best and the worst, it is guaranteed to be close enough to the best to be interesting. These bounds can be generalized to an arbitrary number of buffer cells. If R is the number of buffer cells between cells using the same channel, then the upper and lower bounds on the system capacity are clearly [20]:

$$C_{max} = 1/(R+1) \qquad (1)$$

$$C_{min} = 1/(2R+1) \qquad (2)$$

The upper bound on the system capacity can also be written in terms of the reuse factor $N=R+1$. In this case, $C_{max}=1/N$.

We also posed the following question: What capacity is achieved by random channel placement? That is, cells attempt sequentially to acquire the channel, subject to the reuse constraint. This proceeds until saturation, i.e., until every cell either has a channel or is prohibited from using it. In [20], it is shown that the theoretical capacity is $100\% \times (1-e^{-2})/2 \approx 43.23\%$. Also, in [20], this capacity formula is generalized to arbitrary values of R.

2.2 Elemental Algorithms

2.2.1 Description of Algorithms and Computer Experiments

Now, consider a linear array of cells, as above, with a random starting pattern of 0s (channel not in use in that cell) and 1s (channel in use in that cell). This initial pattern (seed) is randomly generated without adherence to the reuse constraint, $R=1$. At any time, based solely on observation of the current status of adjacent cells, a user in a given cell could spontaneously alter its decision to seize, keep, or relinquish a channel. Obviously, in this scenario, intracell handoffs (rearrangements) are allowed.

We now look to see if there are channel-allocation algorithms that give capacities greater than 40%, i.e., close to the capacity of 50% that an optimal (global) algorithm could achieve, while evolving to adhere to the $R=1$ constraint. It turns out that there *are* such algorithms. We will also highlight which of these algorithms achieve a good capacity with a lower number of updates and will stress those algorithms that appear to make good practical sense. We limit our scope to consideration of temporally and spatially homogeneous algorithms, that is, in each cell, at all times, the algorithm reacts the same way to the same observations. Furthermore, the algorithms we analyze are memoryless, that is, the present decision depends only on the present observation and otherwise makes no use of past decisions or observations.

Four types of algorithms will be studied. The options arise from the following choices: Whether the algorithms in each cell update "one at a time" (asynchronously) or "all at once" (synchronously) and whether the local update algorithms are deterministic or random functions of local observations.

Realisticly, the update timing is neither completely asynchronous nor completely synchronous. While the updating is basically asynchronously driven by call arrivals and completions, an element of synchrony is also present since updates do not occur instantaneously. (Observations take time and an observation time interval for one user can overlap with those of other users.) In the computer experiments, for asynchronous updating, it is assumed that the algorithm in each cell has its update timing randomly clocked by a Poisson process that is statistically independent of the clocks running in the other cells.

We also classify the update algorithms as to whether they are based on deterministic or random functions of the local observations. For deterministic algorithms, for the case of $R=1$, the cell of interest along with its two adjacent neighbors can each be either 0 (channel not in use) or 1 (channel in use). This leads to eight possible observation states. For each possible observation, two possible decisions can be made. Therefore, there are $2^8 = 256$ possible deterministic algorithms. For the probabilistic algorithms, the decisions made in the center cell are no longer binary but rather are weighted by some bias probability for each of the eight possible observations. The bias probabilities can degenerate to 0 or 1 so that the probabilistic algorithms contain the deterministic algorithms as a special class. There are an infinity of probabilistic algorithms; however, by quantizing the probabilities, the number of algorithms becomes finite. If the probabilities are quantized to multiples of $1/4$, there are $5^8 = 390,625$ probabilistic algorithms.

The following notation is useful in discussing the four algorithm types: synchronous deterministic, SD; synchronous probabilistic, $SP(q)$, where q denotes the quantization level; asynchronous deterministic, AD; and asynchronous probabilistic, $AP(q)$, where q is again the quantization level. We represent an algorithm by an eight-vector $[p_{000}, p_{001}, p_{010}, p_{011}, p_{100}, p_{101}, p_{110}, p_{111}]$ where p_{abc} is the probability that b is changed given the local observation (a,b,c). Of course, for deterministic algorithms, this vector has only binary $(0,1)$ entries.

We will be most interested in isotropic algorithms. For them the states 001 and 100 are indistinguishable from each other, as are the states 011 and 110. As a result, there are six instead of eight possible local observations: since $p_{001}=p_{100}$ and $p_{011}=p_{110}$, isotropic algorithms can also be represented by a six-vector $[p_{000}, p_{001}, p_{010}, p_{011}, p_{101}, p_{111}]_{iso}$. Isotropy is a desirable feature of mobile-controlled algorithms since simple measurements are inherently isotropic.

In this study, we try to minimize the number of parameters. Rather than deal with network size, we characterize an infinite linear cellular array by a 100-cell network arranged in a ring. For such a long network, the random starting pattern (seed) should be of minor importance. As we discuss these algorithms, we will check what happens if we greatly increase the number of cells or if we use different starting arrays.

2.2.2 A Simple Example

Before proceeding, consider a 25 cell array as shown in Figure 2.

START STATE

| 1 | 0 | 0 | 0 | 0 | 0 | 0 | 1 | 1 | 1 | 1 | 1 | 1 | 1 | 0 | 1 | 0 | 1 | 0 | 0 | 0 | 1 | 0 | 1 | 1 |

AFTER 25 ITERATIONS (AVERAGE OF 1 UPDATE PER CELL)

| 1 | 0 | 1 | 0 | 1 | 0 | 0 | 1 | 1 | 1 | 1 | 0 | 1 | 1 | 0 | 1 | 0 | 1 | 0 | 1 | 0 | 1 | 0 | 1 | 0 |

AFTER 50 ITERATIONS (2 UPDATES PER CELL)

| 1 | 0 | 1 | 0 | 1 | 0 | 0 | 0 | 1 | 0 | 1 | 0 | 0 | 1 | 0 | 1 | 0 | 1 | 0 | 1 | 0 | 1 | 0 | 1 | 0 |

AFTER 100 ITERATIONS (4 UPDATES PER CELL)

| 1 | 0 | 1 | 0 | 1 | 0 | 1 | 0 | 1 | 0 | 1 | 0 | 0 | 1 | 0 | 1 | 0 | 1 | 0 | 1 | 0 | 1 | 0 | 1 | 0 |

Figure 2: Sample Iterations Of A 25 Cell Array From A Random Start.

We start from a random pattern, with many violations of the reuse constraint, and, using a simple algorithm, demonstrate its self-organizing capabilities. The specific algorithm we are using is one of the *AD* class (cells update "one at a time") in which a user can take a channel only if neither neighboring cell has the channel and, conversely, relinquishes the channel if either neighboring cell is also using it. This is what may be called the *timid* algorithm, since it avoids any and all interference. Keep in mind that, in a more realistic scenario where more than one channel is available for use in the system, relinquishing a channel would usually mean simply switching to another available channel, not dropping the call.* As one can see from the example, after only 25 updates (i.e., an average of 1 update per cell), many of the violations of the reuse

---

* Clearly, users already in the system must be given priority over new users. In other words, the probability of a dropped call should be significantly less than the probability of a blocked call.

constraint have been removed. After 100 updates (an average of 4 per cell), the array is violation-free and has achieved a capacity of 48%.

2.2.3 Results and Discussion

In what follows, computer experiments on *SD*, *SP*(1/4), *AD*, and *AP*(1/4) algorithms are described. These tests were performed to see if we could find some algorithms with the ability to take a network full of violations and evolve it into a violation-free network with only a small degradation from the optimal capacity and with only a small number of updates. (We have already seen an example in the *AD* class.) Each of the algorithms studied was given a statistically independent, randomly generated starting pattern. This was done for a 100-cell array and each algorithm was iterated 1000 times. The evolved array of cells was observed after 10, 100 and 1000 iterations.

2.2.3.1 Synchronous Deterministic Algorithms: We looked at each of the 256 *SD* algorithms to seek out those that produced a violation-free array with capacities of 40% or higher. After 10 iterations there were two such algorithms, by 100 iterations there were eight. After 1000 iterations, only the same eight algorithms achieved capacities greater than 40% and these had no change from what they were after 100 iterations. In interpreting these algorithms, it was clear that only two of the algorithms made any practical sense. For example, one of the impractical algorithms takes the channel even though both adjacent cells are using it and does not take the channel when neither cell is using it. The two algorithms which made some sense were both anisotropic with the following eight-vectors: [1,0,0,1,0,0,0,1] and [1,0,0,0,0,0,1,1]. The former maintains the 110 violation state and the latter maintains the 011 violation state. However, a violation-free state is achieved for the array for both algorithms after only 100 iterations (with a capacity of 44%).

It is of interest to determine if the performance of these algorithms resulted from the particular random start. Each of the eight good algorithms were tested on networks seeded with 100 independent random starts. The results indicated that four of the eight algorithms were flukes, while the other four, including the two above, achieved violation-free states and gave high capacities. The four surviving algorithms were also each given a 1000-cell array to iterate 100 times. The capacities were seen to change little from the 100-cell experiment.

As shown above, the only reasonable *SD* algorithms (i.e., practical and achieving good capacity in a reasonable number of updates) were anisotropic. Only one isotropic algorithm, [1,0,0,1,0,0,1,0], showed up among the top eight competitors and the robustness test showed it was highly volatile: Occasionally it did very well, but it was not a consistent performer. Furthermore, the subtlety of the algorithm may be ill-advised for practical applications since it gives up the channel when it senses that exactly one of its neighbors has it, yet keeps the channel when both neighbors have it. The underlying logic of this algorithm is that by making the interference bad for neighboring cells they might both leave the channel to the center cell.

While we found the two *SD* algorithms of some interest, they were not isotropic and so to implement them may require communication between bases. Keeping to our focus of maximum decentralization, we will not pursue these algorithms further. Our consideration of *SD* algorithms has given us evidence that self-organization is possible but we must look at the other three types of algorithms to find some good, practical algorithms.

2.2.3.2 Synchronous Probabilistic Algorithms: As seen above, for many of the *SD* algorithms, the cells work at counterpurposes to each other and therefore do not achieve a good capacity. For an extreme example, consider the starting condition $111 \cdots 1$. Sensible local decisions (for example, in what we called the *timid* algorithm) give $000 \cdots 0$ and at the next iteration $111 \cdots 1$ again. One way to reduce the effect of cells working at counterpurposes is to introduce probabilistic algorithms (i.e., algorithms that, in effect, flip a possibly biased coin to decide whether or not to take a channel).

Consider the $SP(1/4)$ algorithms. Out of the $5^8 = 390,625$ algorithms that operated on a 100-cell array, hundreds gave a violation-free array with a capacity of 40% or better in less than 100 iterations. Some of these algorithms were isotropic, notably the ones for linear arrays of the form $[1,0,0,0,p,1]_{iso}$. These isotropic algorithms are quite easy to understand since the choices made are natural (notice the fact that a 111 state is immediately changed to a 101). The most interesting aspect is the handling of 011 and 110. In this case, the channel, although interfered with by one interferer, is not necessarily immediately relinquished. There is some "hesitation" to give up the channel; perhaps the interferer will relinquish and the channel can be kept. Notice that this algorithm, which may be termed *aggressive* (although only mildly so), is analogous to having some hysteresis or memory in the handoff process. In the computer experiments, these three algorithms became violation-free within 25 iterations and gave capacities in the 43-44% range. Note that these algorithms do not saturate at perfect capacities (50%) since they do not improve a pair of consecutive 0s straddled by 1s (i.e., a 1001 condition may persist). These three algorithms were tested for consistency of results and for indications of stability in the same manner as with the *SD* algorithms and robustness was confirmed.

2.2.3.3 Asynchronous Deterministic Algorithms: Although, in a given situation, several mobile or portable units, given the practical aspects of updating, may change at essentially the same time, the situation in which all cells update at each time instant is not realistic. In this subsection and the next, we consider the other extreme.

The 256 deterministic algorithms were retested, but under the assumption that the algorithm in each cell is run with a random independent (asynchronous) clock. Again, 100 cells were assumed and observations were made at a *mean* value of 10, 100 and 1000 updates *per cell* (to match the *SD* case). (We should emphasize that, for the asynchronous algorithms, by one iteration we mean that the expected number of updates per cell is one.) The list of algorithms giving violation-free capacities above 40% expanded as the number of updates per cell increased. There were a total of 7 algorithms at 10 updates, 8 algorithms at 100 updates, and 20 algorithms at 1000 updates. After 1000 updates, 13 of the 20 algorithms achieved the maximum capacity of 50%. If we inspect the evolving array after each iteration, we see that the number of cells with channels drops at first to clear violations and then rises back up as cells get channels without incurring violations. Clearly, in a more realistic situation, where the system does not start in a state full of violations, the number of iterations required to achieve a violation-free state will be much smaller.

While most of the 20 algorithms were anisotropic, there were some isotropic ones as well. For example, at 10, 100 and 1000 updates per cell, the isotropic algorithms [1,0,0,1,0,0,1,1] and [1,0,0,1,0,0,1,0] appeared on the list. At 1000 updates, [1,1,0,1,1,0,1,0], which is an aggressive isotropic algorithm, appeared with a perfect capacity of 50%. However, the last two algorithms are not practical since a 111 state is maintained. The most sensible of the isotropic algorithms [1,0,0,1,0,0,1,1], which we previously termed the *timid* algorithm, was given 100 additional random starting patterns to test for robustness. After an average of 1000 updates per cell, the capacities ranged from 40% to 48%. All 100 outcomes were violation-free.

2.2.3.4 Asynchronous Probabilistic Algorithms: Probabilistic algorithms still make sense even though we have found deterministic, isotropic algorithms that work. This is because some randomness (as by possibly taking a channel even though there is interference) may reduce blocking at the expense of a simple rearrangement. In addition, a probabilistic algorithm may be used to speed up channel acquisition.

There are many effective (practical) isotropic algorithms of the *AP* type. Some of these algorithms are isotropic, notably the ones for linear arrays of the form $[1,p, 0,q, 0,1]_{iso}$. Recall the meaning of the notation:

LOCAL OBS.	DECISION
000	take channel with probability 1
001 or 100	take channel with probability p
010	no change
011 or 110	release channel with probability q
101	no change
111	release channel with probability 1

The effectiveness of certain AP(1/4) algorithms operating in a randomly seeded 100 cell network is shown below:

SAMPLE CAPACITIES FOR ISOTROPIC ALGORITHMS
OF THE FORM $[1,p,0,q,0,1]_{iso}$

Probability of Change		Average Number of Updates per Cell		
$p_{001\to011}$	$q_{011\to001}$	10	100	1000
0	1/4	43	43	43
0	1/2	45	45	45
0	3/4	43	43	43
0	1	44	44	44
1/4	1/2	-	-	50
1/4	3/4	-	47	50
1/4	1	-	49	50
1/2	1/4	-	-	50
1/2	1/2	-	49	50
1/2	3/4	-	-	50
3/4	1/2	-	-	50
3/4	3/4	-	-	50
3/4	1	-	50	50
1	1/4	-	-	50
1	1/2	-	-	50
1	3/4	-	-	50
1	1	-	50	50

The "-" designation in the table means that the evolved state had violations. Notice that all of these algorithms involve fairly natural choices. By varying p and q, the algorithms range from the most timid ($p=0$, $q=1$) through some aggressive (but reasonable) algorithms. Performance robustness was also demonstrated using the method previously discussed.

As can be seen from the table, the first four algorithms are timid-type algorithms ($p=0$) with varying degrees of "hesitation" (up to giving up the channel "immediately", $q=1$). Each of these algorithms converge to a violation-free state very quickly with good capacities. The rest of the algorithms are of the aggressive type and, while they take longer to converge, they eventually converge to the perfect state (50%). We can interpret this to mean that we can trade off the number of reconfigurations per call for increased capacity (or, improved blocking probability) by using an aggressive algorithm. It is a concern, however, that, by taking the channel even though it is being used nearby, a user may force a cochannel user to look for another channel. In a heavily loaded system, an unstable situation may result with an unreasonable number of reconfigurations per call. We will address this question in more detail in the future. However, one possibility is to vary the bias probability (or, equivalently, adapt the interference thresholds) depending on the traffic load and propagation conditions.

2.2.4 Summary and Conclusions

In this section, a linear array of cells was operated from random starts. At any time, based solely on the observation of the current status of a cell and its adjacent cells, a user in a given cell could spontaneously alter its decision to seize, keep, or relinquish a channel. We considered cases where the cells could rearrange asynchronously (one at a time) or synchronously (all at once). Using simulation, we explored *all* algorithms for this scenario (both deterministic and, within a quantization of 1/4, random). Several interesting conclusions (which should apply in more realistic cases) can be drawn from the consideration of these elemental algorithms: (1) Systems can self-organize with little loss in capacity by using simple algorithms where decisions are based solely on local observations. (2) Using the system capacity as the performance measure of interest, the intuitively best algorithms turn out to *be* among the best. (3) Call blocking can be reduced, in some cases, with only a small increase in the number of reconfigurations required, by seizing a channel even though there is interference nearby.

3. Area Microcells

3.1 Bounds on System Capacity

Consider a planar array of hexagonal cells, as shown in Figure 3, with only a single channel available for use. Also, assume that, if the channel is used in a given cell, it cannot be used in a ring of cells around that cell.

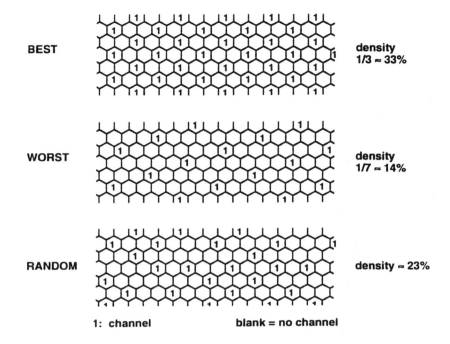

BEST   density 1/3 ≈ 33%

WORST   density 1/7 ≈ 14%

RANDOM   density ≈ 23%

1: channel   blank = no channel

Figure 3: Planar Hexagonal Arrays. This model can be used for urban and suburban areas and outdoor shopping malls. Best capacity, C_{max}=33%, Worst capacity, C_{min}=14%, and Random capacity C_{ran} ≈ 23%.

It is easy to show that, if the cells are *globally* coordinated, the best arrangement is for the channel to be reused in clusters of three cells, yielding a capacity (or efficiency), C_{max}=1/3 (33%). In the worst case, the channel is reused in a cluster of seven cells, yielding a capacity, C_{min}=1/7 (14%). These bounds can be generalized for an arbitrary number of rings, R, as [20]

$$C_{max} = \begin{cases} \dfrac{4}{3(R+1)^2+1} & R \ even \\[3mm] \dfrac{4}{3(R+1)^2} & R \ odd \end{cases} \qquad (3)$$

As before, we can relate this to the reuse factor N as C_{max}=1/N, where $N=i^2+ij+j^2$ and i and j are integers [2]. For R even, $i=j=(R+1)/2$ and, for R odd, $i=R/2$ and $j=R/2+1$. Furthermore, it can also be shown that

[20]

$$C_{min} = \frac{1}{3R(R+1)+1} \tag{4}$$

Random placement, estimated via simulations, gives 23%. Notice that this is a much larger penalty (31%) than in the linear case (14%) when compared to the best capacity. In this case, the "aggressive" algorithm may be essential.

3.2 Elemental Algorithms

3.2.1 Description of Algorithms
Based on our previous experience with the linear array of cells, for the hexagonal array of cells, we restricted our study to algorithms which update asynchronously ("one at a time") and which are isotropic. Both deterministic and probabilistic approaches were studied. We concentrated on practical algorithms of both the timid and aggressive types.

We considered a 24×22 array of cells comprising a torus in two dimensions. The array is randomly seeded. The algorithms operate by totaling the number of cells in a ring of six around the cell of interest that currently use the channel and then comparing the number to a threshold to determine whether to seize, keep, or relinquish the channel. In particular, algorithms will be described by a four-vector $[t_0,t_1,p_0,p_1]$. The first two parameters are threshold values: t_0 indicates the number of interferers for which a user cannot take a channel if it does not already have it; t_1 indicates the number of interferers for which a user must release a channel. The last two parameters indicate the probability that if the number of interferers is less than t_0 (t_1), the user takes (releases) the channel. In addition, if there are no interferers, the user takes the channel with probability 1 or, if the user is already using the channel, does not relinquish it.

3.2.2 Results and Discussion
We illustrate the performance of these algorithms with a few examples. The first two examples are algorithms of the timid class: $[1,1,1,1]$ which is the most timid, avoiding any and all interference, and $[1,2,1,0.5]$, which hesitates in giving up the channel when interference presents itself. The inclusion of hysteresis in the handoff process (in this case, modeled by some probability of not giving up the channel) will be essential in any realistic control algorithm. From a random start (full of violations), the $[1,1,1,1]$ algorithm converged to a violation-free state after only 20 updates per cell. The $[1,2,1,0.5]$ algorithm took 40 updates per cell. Each algorithm achieved a capacity of 23%. Robustness was tested by

giving each algorithm 50 independent random starts. For all 50 starts, both algorithms converged to violation-free states with capacities of 23%.

The last two examples are algorithms of the aggressive class: [2,1,1,1], which takes the channel if there is one interferer, but releases the channel if the interferer is there during the next update, and [2,2,0.5,0.5], which includes aspects of aggression in taking a channel and hesitation in giving it up. From a random start, both algorithms converged to a perfect state (33%), but required 150 updates per cell to reach this state. Again, we see that we can achieve the best possible capacity using these algorithms if we are willing to accept an increase in the number of reconfigurations per call. When the algorithms were tested on 50 random starts, most of the time they converged to a violation-free state with capacities of 33%. However, in some cases, a small number of violations persisted even when a very large number of updates were used. When traffic considerations are included, the issue of "stability" of reconfigurations is a concern. Section 5 includes an initial look at this issue.

4. Simulated Annealing Interpretation Of Results

The algorithms we have been treating are memoryless and isotropic. They have all also been spatially and temporally homogeneous to the extreme. As we have mentioned, because of this simplicity of structure they are easily enumerated. Figure 4 summarizes the number of algorithms of various types.

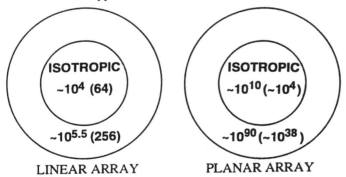

LINEAR ARRAY   PLANAR ARRAY

Figure 4: Enumeration Of Elemental Algorithms. The two larger circles represent anisotropic as well as isotropic elemental algorithms. The number of probabilistic algorithms assumes a quantization of 1/4, so their number is $5^{no.\ obs.}$. The number of deterministic algorithms, $2^{no.\ obs.}$, is shown in parentheses.

As mentioned in Section 3 we explored all such linear algorithms. In Section 4, we reported an investigation of a very broad class of isotropic planar algorithms. Such elemental distributed autonomous DCA algorithms represent a very high dimensional dynamic for evolving cellular arrays. Certain overall properties of these algorithm classes, that we have observed in our investigations, are conveniently summarized using the well known simulated annealing analog [21]. This is depicted below where we have taken the liberty of a figurative description in the plane.

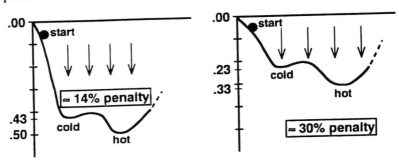

Figure 5: Simulated Annealing

To illustrate simulated annealing two examples of a particle moving on a terrain are shown. In both cases, the particle appears headed for the limiting position labeled "cold". If somehow the motion can be perturbed, it might make it over the hump to come to rest at the limiting position labeled "hot". To achieve the hot limit instead of the cold limit perhaps some noisy effect can be added to the force or some tremor of the terrain could be induced, or maybe some activity induced in the particle itself. In the basin of attraction of the hot limit the perturbative influence would turn off. Increasing the perturbative influence is referred to as "raising the temperature", a loose terminology borrowed from materials science where raising the temperature (followed by cooling) is used in improving the quality of glasses and metals.

The rough analog covering our findings goes as follows. The particle corresponds to the channel populated array when the array is assumed to be very large. The forces acting on it correspond to the distributed DCA algorithm. The temperature corresponds to reconfigurations per call and the ordinate extrema to limiting channel population densities. The left part of the figure is for linear arrays and the right for planar arrays. What we have observed in our investigations of vast classes of algorithms, for both linear and planar arrays are basically three types of algorithms. The first type covers the overwhelming majority of algorithms which we observe to be worthless for evolving the array to a good channel density. By a good density we mean a channel density of over 40% out of a maximum of 50% for a linear array and say over 20% out of 33 1/3% for a planar array. These "worthless" algorithms either do not converge or converge to poor channel densities and/or to arrays with violations. There are rare good algorithms of two types. There are the timid type that go to the cold limit. And there are the aggressive type that converge to the hot limit at the expense of reconfiguring channel assignments before coming to rest.

5. Multichannel Algorithms With Traffic

5.1 From Elemental Algorithms to Multilevel Algorithms With Traffic

Next we turn to multichannel algorithms with traffic. We assume Poisson arrivals and exponential call durations and a uniform load of ρ Erlangs in each cell. It is useful to consider multichannel algorithms as integrating elemental algorithms: one elemental algorithm per channel.

The focus in contrasting multichannel algorithms with traffic is the accommodation of spatially and temporally localized strong demand for channels. The main question is: How do such algorithms compare in handling situations where there is a risk of blocking calls? So far, in our testing of elemental algorithms we have not included traffic considerations. However, we have been addressing the capability of elemental algorithms to saturate an array with interference free copies of a single channel. So our study of elemental algorithms has been focusing on just the "push comes to shove" type of situation that is of interest for multichannel algorithms. Indeed, our elemental investigations can be viewed as tacitly assuming an insatiable demand of one usable channel for each cell, and probing the comparative ability of algorithms, that when confronted with such strong demand, to preferably access a channel, but otherwise facilitate a neighbor's gaining access.

5.2 Five Key DCA Algorithms

Five key multichannel DCA algorithms are listed below in order of increased aggressiveness. The first four listed do not permit dropped calls. The first three are completely decentralized in that they involve actions based only on measurements of interference from neighboring cells.

TIMID: A channel is accessed if and only if there is no interference. If an originating call cannot be so accessed it is blocked. No reconfigurations of channel assignments are allowed with this algorithm so stability is not a problem.

POLITE AGGRESSIVE (PA): If TIMID does not give access, try to get access by reconfiguration of one randomly chosen active user in an adjoining cell (to limit disruption, the random choice is only among the single interferer cochannel users) In seeking to reconfigure the imposed upon user acts in a timid manner so there is no cascade of impositions to reconfigure.

PERSISTENT POLITE AGGRESSIVE: (PPA): Same as PA, except all (single interferer) neighbors can be interrupted to see if a reconfiguration of one of them is possible so the originating call can be accepted. As with PA, any imposed upon user acts in a timid manner when seeking to reconfigure.

The aggressive action in the previous two algorithms, as well as the response to it by the intruded upon user, is all conducted in an automatic fashion through the use of local measurements of cochannel interference. These two aggressive algorithms involve at most one reconfiguration per originating call. Consequently, aggressiveness is integrated into these two algorithms in a manner that assures stability.

MAXIMUM PACKING (MP): Channel access if and only if there exists a reconfiguration that permits it. This algorithm requires global information and centralized control. MP is a well known algorithm and in the planar array case it is known to be NP-complete [24]. We do not consider MP as a practical algorithm. As we shall see it is included on this list for comparison purposes.

For the fifth algorithm we need to define the notion of an unsuccessful call. It is one that is dropped before completion or blocked on attempting entry. (We emphasize that the first four algorithms above do not ever involve such dropped calls, so for them, an unsuccessful call is a simply a call blocked when attempting entry.)

OPTIMUM AGGRESSIVE: The algorithm achieving the minimum probability of an unsuccessful call. A constructive way of defining this algorithm is unknown to the authors.

5.3 Comparative Performance Of Algorithms

For the sample case of 36 channels, the comparative performance of algorithms is shown in Figure 6.

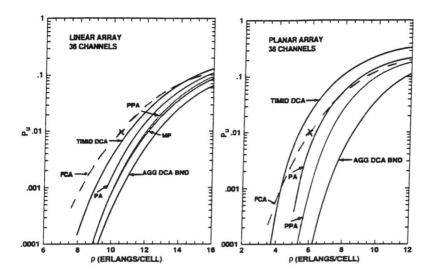

Figure 6: Comparison Of DCA Algorithms, Probability of Unsuccessful Call Versus ρ. (a) Linear Array, (b) Planar Array

For these two figures, the elemental algorithm saturation densities are used in deriving the TIMID performance characteristics as well as for determining the lower bound curves shown for the optimum aggressive algorithm in Figures 6a and 6b. See [23] for an explanation of how the saturation densities, that we obtained earlier for elemental algorithms, were used to obtain these two curves, as well as a discussion of how the other curves were obtained.

The fixed channel allocation, FCA, performance at the important 1% blocking level is marked as X on both figures. We note that in both cases it is comparable in blocking performance to the TIMID algorithm. The slight shortfall in the planar case is inconsequential and is seen to be easily compensated with a small level of aggressiveness. FCA is associated with centralized control, an option which appears to be impractical for microcells.

Figure 6a suggests that PPA and MP are asymptotically equivalent as the loading, ρ, lightens. (We showed this analytically in a companion paper [26].) At the 1% blocking level, by comparing with MP performance, we see that it is pointless to seek to improve on PPA performance if one does not permit the turning away of calls when channels for them are available†. Furthermore, at such low blocking, whether one can improve over PPA or MP by such turning away seems questionable.

The PA curve in both figures 6a and 6b is seen to be disposed roughly midway between TIMID and PPA. So in terms of the number of neighbors "asked" for help, diminishing returns is seen to set in rather quickly: Just imposing on one of the single interferer neighbors to reconfigure if possible is seen to go a long way.

We note that for the planar case the MP curve is not shown as it was not known how to effectively compute it. The similar disposition of curves in the collection of performance characteristics of the four non-MP algorithms in 6a with their four counterparts in 6b leads us to conjecture that the MP curve is again very close to the PPA curve (for low P_u) in the planar case as it was in the linear case. Adding credence to this conjecture, we have, in the planar case, as with the linear array case, proved asymptotic equivalence of the PPA and MP performance characteristics as ρ decreases toward zero. This is reported in [26].

Figures 6a and 6b show that by being more aggressive one can reduce call blocking, or at the same blocking level one can accept a larger demand. A third way to use increased aggressiveness to advantage, is to provide for the same demand at the same blocking probability but with a reduced number of channels: In [23] it is shown that PPA offers a 10% reduction in the number of channels over TIMID for linear arrays and a 19% reduction for planar arrays.

6. Conclusions

We have studied the capacity of linear and area microcells using dynamic channel allocation with the mobile or portable making autonomous decisions based only on local observations. We demonstrated that microcellular systems can self-organize, with little loss in capacity, by using channel-allocation algorithms that are simple, practical, and local. We calculated performance for both deterministic and probabilistic algorithms. Two classes of isotropic algorithms look particularly promising: The timid class, which is the simplest, and the aggressive class, which could provide improvements in system capacity and blocking

---

† For sufficiently heavy traffic, one can reduce the probability of blocking, by blocking certain calls when there are channels available [25].

probability. In particular, at the expense of additional reconfigurations (or, intracell handoffs) per call, these "local" algorithms can approach the capacity achieved by a global channel-allocation strategy. The additional cost in terms of a potential increase in the call-dropping probability needs to be investigated. The simulated annealing concept provided a convenient way of interpreting our findings. In Section 5 we took an initial look at multilevel algorithms with traffic considerations included. We saw that increasing levels of aggressiveness can be included to obtain performance benefits in a stable manner.

REFERENCES

[1]   V. H. MacDonald, "The Cellular Concept", *Bell Sys. Tech. J.*, Vol. 58, No. 1, Part 3, Jan. 1979, pp. 15-41.

[2]   S. T. S. Chia, R. Steele, E. Green, and A. Baran, "Propagation and BER Measurements for a Microcellular System", *J. of IERE*, Vol. 57, No. 6, Nov./Dec. 1987, pp. S255-S266.

[3]   D. A. McFarlane and S. T. S. Chia, "Microcellular Mobile Radio Systems", *Br. Telecom. Tech. J.*, Jan. 1990, pp. 79-84.

[4]   F. Lotse and A. Wejke, "Propagation Measurements for Microcells in Central Stockholm", *Proc. of VTC'90*, pp. 539-541.

[5]   N. Amitay, L. J. Greenstein, and G. J. Owens, "Measurement-Based Estimates of BER Performance in Urban LOS Microcells at 900 MHz", *Proc. of VTC'91*, pp. 904-909.

[6]   A. J. Rustako, Jr., N. Amitay, G. J. Owens, and R. S. Roman, "Radio Propagation Measurements at Microwave Frequencies for LOS Microcellular Mobile and Personal Communications", *IEEE Trans. on Veh. Tech.*, Vol. 40, No. 1, Feb. 1991, pp. 203-210.

[7]   R. Beck and H. Panzer, "Strategies for Handover and DCA in Micro-Cellular Mobile Radio Systems", *"Proc. of VTC'89"*, pp. 178-185.

[8]   H. Panzer and R. Beck, "Adaptive Resource Allocation in Metropolitan Area Cellular Mobile Radio Systems", *Proc. of VTC'90*, pp. 638-645.

[9]   I. Horikawa and M. Hirono, "A Multi-Carrier Switching TDMA-TDD Microcell Telecommunications System", *Proc. of VTC'90*, pp. 167-171.

[10] M. Yokoyama, "Decentralization and Distribution in Network Control of Mobile Radio Systems", *Trans. of IECICE*, Vol. E 73, No. 10, Oct. 1990, pp. 1579-1586.

[11] D. Akerberg, "Properties of a TDMA Picocellular Office Communication System", *Proc. of VTC'89*, pp. 186-191.

[12] C. Buckingham, G. K. Wolterink, and D. Akerberg, "A Business Cordless PABX Telephone System on 800 MHZ Based on the DECT Technology", *IEEE Commun. Mag.*, Jan. 1991, pp. 105-110.

[13] J. C.-I. Chuang, "Autonomous Frequency Assignment and Access for TDMA Personal Portable Radio Communications", *Proc. of VTC'91*, pp. 699-703.

[14] B. Gudmundson and O. Grimlund, "Handoff in Microcellular Based Personal Telephone Systems", *Second Rutgers Workshop on Third Generation Wireless Information Networks*, Oct. 18-19, 1990, pp. 160-171.

[15] C.-L. I, L. J. Greenstein, and R. D. Gitlin, "A Microcell/Macrocell Architecture for Low- and High-Mobility Wireless Users", *Proc. GLOBECOM '92*, submitted to IEEE J. on Sel. Areas in Comm.

[16] S. T. S. Chia and R. J. Warburton, "Handover Criteria for City Microcellular Radio Systems", *Proc. of VTC'90*, pp. 276-281.

[17] D. C. Cox and D. O. Reudink, "Increasing Channel Occupancy in Large-Scale Mobile Radio Systems: Dynamic Channel Reassignment", *IEEE Trans. on Commun.*, Vol. COM-21, No. 11, Nov. 1973, pp. 1302-1306.

[18] P.-A. Raymond, "Performance Analysis of Cellular Networks", *IEEE Trans. on Commun.*, Vol. 39, No. 12, Dec. 1991, pp. 1787-1793

[19] L. J. Cimini, Jr. and G. J. Foschini, "Distributed Algorithms For DCA In Microcellular Systems", *Proc. of VTC'92*, pa.p 641-644.

[20] L. J. Cimini, Jr., G. J. Foschini, and L. A. Shepp, "Single-Channel User-Capacity Calculations For Self-Organizing Cellular Systems", submitted to *IEEE Trans. on Commun.*

[21] W. H. Press, B. P. Flannery, S. A. Teukolsky and W. T. Vetterling, *Numerical Recipes,* Cambridge U. Press, New York, 1986, Chapter 10.

[22] R. A. Valenzuela, "Dynamic Resource Allocation in Line-Of-Sight Microcells", to appear in *Proc. of GLOBECOM '92*

[23] L. J. Cimini, Jr., G. J. Foschini and C.-L. I, "Call Blocking Performance of Distributed Algorithms for DCA in Microcells", *Proc. of ICC'92*.

[24] K. Keeler, "On Maximum-Packing Strategy for Channel Assignment in Cellular Systems", submitted to *IEEE Trans. on Commun.*

[25] C.-L. I, "Distributed DCA Algorithms in Microcells Under Light Traffic Loading", submitted for ICC'93.

[26] F. P. Kelly, "Stochastic Models of Computer Communication Systems", *J. R. Statist. Soc. B*, (1985) Vol. 47, No. 3, pp. 379-395.

ADAPTIVE CHANNEL ALLOCATION IN CELLULAR NETWORKS WITH MULTI-USER PLATFORMS

Jelena F. Vucetic, Dragomir D. Dimitrijevic
GTE Laboratories Inc.

Abstract - This paper proposes a hardware solution to the efficient utilization of cellular networks with single- and multi-terminal platforms. In such networks, a mobile platform (e.g., an airplane) can carry more than one wireless terminal. A good utilization of available channels as a shared resource is important for quality and efficient communications in the network. In this paper, we propose the Integrated Channel Manager (ICM), an architecture for fast adaptive channel allocation. It is an integrated controller connected to the system bus within the network switch. Its main advantage is a fast allocation of available channels when a request for a call initialization or a hand-off exists. Its efficiency is achieved via channel allocation functions supported by a hardware with high degree of parallelism. The ICM supports both single and multiple hand-offs. It allows an efficient rejection of a call when the call cannot be supported. Thus, it reduces the processing overhead for rejected calls.

1. INTRODUCTION

Cellular networks provide efficient wireless communications between mobile users. In achieving this goal, a trade-off must be made between the area coverage and the user capacity. For this reason, Eriksson [Eri88] proposed and analyzed an *adaptive channel allocation,* where each channel in the available radio bandwidth may be used by any base station in the network, as long as a certain level of carrier to interferer ratio is maintained.

There are three types of stations in a cellular network: a switch, base stations and mobile platforms. Base stations are connected with the network switch by a fixed infrastructure. Each base station controls mobile platforms in its covering area called a "cell". We assume that cells have a hexagonal form. Mobile platforms carry one or more wireless terminals while roaming through the area serviced by the network. When calls on a platform can be handled better by another base station in another cell (e.g., better signal to noise ratio), the platform is registered (or handed-off) to the new base station. If a call on the platform is in progress, a new channel in the new cell must be assigned to the call in progress. For a successful hand-off, such radio channel must be available in the new cell. Otherwise, the call is rejected. This task is performed by the cellular network switch. The call is interrupted (blocked) during a hand-off execution. This deteriorates the quality of a transmission. Hence, the hand-off time must be reduced using fast channel allocation [Rap90].

Nanda and Goodman [Nan90] propose and evaluate a Dynamic Resource Acquisition (DRA) distributed algorithm in terms of transmission quality, spectrum utilization, and cost. In [Nan91], the DRA algorithm is applied to a microcellular third-generation network to accommodate large traffic variations, assuming that each radio carrier is capable of carrying several voice channels simultaneously. Rappaport [Rap90] introduces the problem *multiple hand-offs* where platforms may carry more than one wireless terminals (e.g., airplanes). This increases the requirements for efficiency of the switch and channel allocation algorithms. Other published work on this topic may be found in [Vuc92a-b].

In this paper, we propose an adaptive channel allocation scheme suitable for implementation in hardware. The implementation represents an original special-purpose VLSI component called *Integrated Channel Manager* (ICM). The ICM can support a set of adaptive channel allocation functions [Vuc92a-b]. Its architecture provides a high level of parallelism in the execution and a large internal data base. Its main advantage is a *fast search* of available channels subject to the co-channel and adjacent-channel interference constraints. Also, the ICM provides *efficient* management of the data base containing the information on the network (e.g., channel availability, current channel assignments, locations of platforms, etc.). These advantages are achieved using a fast memory access mechanism called *marking access* [Ale86, Vuc86, Vuc90, Vuc91] within the ICM. The marking access mechanism may be implemented in hardware with a high degree of parallelism which additionally improves the efficiency.

The paper is organized as follows. A model of a cellular network is described in Section 2. Channel allocation functions, related data structures, a search mechanism, and a channel allocation algorithm are defined in Section 3. In Section 4, we describe the ICM architecture and its implementation in hardware. The performance of a network (throughput and speed) controlled by the ICM is analyzed in Section 5. Section 6 gives a conclusion.

2. A MODEL OF A CELLULAR NETWORK

In this section we describe a model of a cellular network. We define various parameters and restrictions used in our analysis.

2.1. Network topology and configuration: In our model, a cellular network consists of three types of stations: a switch, base stations and mobile platforms. The switch and base stations are connected by a fixed-topology infrastructure. Assume that the switch can support up to S base stations. The mobile platforms move through the area serviced by the network. They are temporarily registered to the base stations corresponding to cells in which platforms are currently located. We assume that a base station is located in the center of a hexagonal cell and mobile platforms are uniformly distributed within a cell. The switch can support up to F platforms simultaneously. A mobile platform can handle up to L wireless terminals at a time. Each base station in the network can support up to N_{max} wireless terminals simultaneously.

2.2. Traffic requirements: We assume that the average total call initialization rate in the network is λ. The average duration of a call is denoted as τ. The average number of hand-offs during a call in progress is H. This parameter depends on the size of a cell, average speed of wireless terminals, and average call duration. We assume that all necessary inputs, such as the network requirement matrix, the list of radio neighbors for each base station, and moving directions of mobile platforms are defined and available.

2.3 Available resources: Wireless terminal and its base station communicate via a radio channel. The available radio bandwidth consists of K channels. The network switch allocates temporarily an available channel (if it exists) to the terminal which wants to communicate. For more efficient use of the bandwidth, a frequency-reuse concept is proposed [Lee89, Par89]. The cells are grouped into *clusters* so that all channels in the available bandwidth can be assigned to a cluster. The frequency reuse concept allows that the same channels can be used in different cells of the network simultaneously as long as the carrier-to-interference ratio is acceptable. According to the state-of-the-art technology, it is estimated [Par89] that at a seven-cell cluster is generally a good trade-off between the area coverage and the user capacity. The network switch controls channel allocation [Eri88, Good90, Kum90].

2.4. Restrictions: In the adaptive channel allocation concept, several restrictions should be provided to reduce the impact of *co-channel* and *adjacent-channel* *interference*. Co-channel interference appears in a receiver of any terminal operating on any channel. It is caused by signals from another terminals using the same channel. To reduce this type of interference, the same channel should not be used more than once within a cell or in adjacent cells. Adjacent-channel interference also deteriorates the quality of transmission. It occurs in a receiver of any terminal operating on any channel in presence of a high-level signal from the adjacent radio channel. To reduce the influence of adjacent-channel interference on the quality of transmission, adjacent channels should not be used simultaneously in the same cell. As in [Par89], we assume that the adjacent-channel interference, which occurs when adjacent channels are used in adjacent cells, can be neglected. For simplicity of the analysis, we assume that the communication should be established between two mobile users only. With this assumption, two radio channels must be available for each call: one for the calling and another for the called party.

3. CHANNEL ALLOCATION FUNCTIONS

Radio bandwidth in cellular network is a limited resource shared by all users. Since the number of mobile users and traffic demands grow rapidly, proper utilization of the radio bandwidth is critical for network performance. In this analysis, we propose an efficient adaptive channel allocation algorithm which is executed in the network switch. The algorithm is based on data structures which have the form of, so called, an index vector. The index vector is a logical structure which enables fast and easy access to specified data in a database when the location of the data in the database is unknown, and/or the existence of the data in the database is uncertain.

3.1 Index Vector as a Generic Data Structure: Channel allocation functions are used by the network switch to maintain the information on the network and efficiently utilize the limited radio bandwidth. These functions include the following activities:

- Search for an available channel subject to the interference constraints;
- Maintenance and update of information on availability of all the channels;
- Maintenance and update of information on all channel-to-terminal assignments.

For efficient channel allocation, a storage for information on channel availability and assignment, and fast access to and/or update of the information in the storage are necessary. For that purpose, the ICM uses a generic data structure called *Index Vector (IV)*, and a memory access mechanism called *Marking Access* [Ale86, Vuc86, Vuc90]. An *IV* of degree N is a vector consisting of 2^N binary elements. An element $IV[A]$ is *marked* if its content is 1. If the content of $IV[A]$ is 0, the element is *unmarked*. In the following text, we shall call this vector the *main IV*, to distinguish it from the auxiliary *IV*s which will be defined in Section 4.

Marking access is a mechanism for access to the index vector. It speeds up searching for certain data in large data bases when data locations are not known and their existence is uncertain. Marking access consists of four operations which may be performed on an *IV*:

- *mark* writes 1 into an *IV* element.
- *read* determines if an *IV* element is marked;
- *scan* lists all marked *IV* elements;
- *delete* unmarks one or several *IV* elements.

In Section 3, we shall use these operations as generic operations, i.e. we shall not address the details of a particular implementation. Later, in Section 4, we shall propose a fast implementation as an integrated VLSI chip.

3.2. Data Structures of the ICM: The ICM algorithm for adaptive channel allocation is based on the following data structures:

Index of Clusters (IC) stores information on the availability of a channel in a cluster. Its capacity is $P \times K$ bits, where P is the number of clusters in the network, and K is the number of channels in the bandwidth. Note that $P \geq \lceil S/7 \rceil$. The IC is a set of P main index vectors with K elements each. Marking access to such a data structure is used to find an available channel. $IC[p,k]$ defines the status of channel k in cluster p. If this element is marked, the channel k is available in cluster p. Otherwise, the in channel is busy.

Index of Base Stations (IBS) stores information on the channels used by base stations. Its capacity is $S \times K$ bits. An IBS is a set of S main index vectors each comprising of K elements. Each vector refers to a unique base station in the network. Each bit-element of an index vector corresponds uniquely to a channel. An element $IBS[s,k]$ shows the status of channel k within station s. If $IBS[s,k]$ is unmarked, channel k is used in base station s. Otherwise, the channel is available in base station s.

Vector of Current Numbers of active terminals (VCN) stores information on the current number of wireless terminals supported by a base station. The VCN has S elements. If NC_s wireless terminals are currently active in a cell controlled by base station s, then $VCN[s] = NC_s$. Note that $NC_s \leq N_{max}$.

Table of Neighbors (TN) contains information on all radio neighbor base stations for each base station in the network. *Radio neighbors* are stations which may suffer from interference. We also assume that a base station is its own radio neighbor. It is possible, but not necessary, that two stations are geographical and radio neighbors at the same time. The TN has $S \times S$ bits organized as S main index vectors with S bit-elements each. If $TN[s,t]$ is marked, base stations s and t are radio neighbors $(0 < s, t \leq S)$. Therefore, they should not use the same/adjacent channels simultaneously.

Table of Platforms (TP) keeps track of positions of mobile terminals. Its capacity is F locations, where F is the maximal number of platforms in the network. If $TP[f]=s$, platform f is currently in the cell covered by base station s, where $1 \leq f \leq F$, and $1 \leq s \leq S$.

Index of Active Labels (IAL) contains information on channels and unique labels assigned to active terminals for the duration of a call. The IAL has capacity of $S \times K \times N_{max}$ bits. N_{max} denotes the maximal number of active terminals in a base station. If $IAL[s,k,n]$ is unmarked, channel k is used within base station s by wireless terminal which current label is n, $1 \leq n \leq N_{max}$.

Index of Platforms (IP) and *Table of Terminals (TT)* contain information on the labels currently assigned to wireless terminals. The IP and TT have $F \times N_{max}$ bits and $F \times N_{max}$ integers, respectively. When terminal t on platform f becomes active, label n is assigned to it. Also, $IP[f,n]$ is unmarked and $TT[f,n]$ is set to t. When the terminal terminates the call, the corresponding label n is released, i. e. $IP[f,n]$ is marked and $TT[f,n]=0$.

Table of Base stations (TB) and *Table of Labels (TL)* store information on pairs of base stations and assigned labels related to a call. Both TB and TL consist of $S \times N_{max}$ locations each. Let us assume that terminals t_a and t_b are involved in a call. Also, let us assume that terminals t_a and t_b are currently under control of base stations a and b, and they are assigned labels n_a and n_b, respectively. Then, $TB[a,n_a] = b$, $TL[a,n_a] = n_b$, $TB[b,n_b] = a$, and $TL[b,n_b] = n_a$.

Vector of Current Platform Orders (VCPO) contains information on the current number of wireless terminals carried by a platform. $VCPO[f]=l_f$ means that platform f currently supports l_f active terminals, where $l_f \leq L$.

3.3 Channel Allocation Functions: Using the data structures listed above, the ICM is designed to execute several functions which define adaptive channel allocation [Vuc91, Vuc92a-b]. The ICM functions can be divided into two general classes: *a*) Initial Mode functions; *b*) Nominal Mode functions.

The Initial Mode functions are used to set up the initial network operating conditions. These functions are: Reset, System Initialization, and Registration. The Nominal Mode functions are used for call processing. These functions are: Call Initialization, Call Termination, and Hand-Off. In this paper, we shall describe only the Nominal Mode functions.

Call Initialization:

Assumptions: Terminal t_a on platform f_a wants to call terminal t_b on platform f_b, where $1 \le f_a, f_b \le F$. Platforms f_a and f_b are currently under control of base stations a and b, respectively, where $1 \le a, b \le S$. Platforms f_a and f_b currently support l_f and l_{f_b} active terminals, respectively. Base stations a and b belong to clusters p_a and p_b, respectively. Terminals t_a and t_b are not busy when the call is initialized, so an eventual rejection of the call may occur only due to the unavailability of channels, not because the called terminal is busy.

1. Check if the capacity of base stations are reached: If $N_{max} = VCN[a]$ and/or $N_{max} = VCN[b]$, reject the call and terminate the procedure.
2. Check if the capacity of the corresponding platforms are reached: If $L = l_{f_b}$ and/or $L = l_{f_a}$, reject the call and terminate the procedure.
3. Determine channels k_a and k_b for base stations a and b, respectively, subject to the interference constraints: Scan $IC[p_a, k_a]$ and $IC[p_b, k_b]$ for marked k_a and k_b, respectively; If such channels do not exist, reject the call and terminate the procedure.
4. Check co-channel interference for channels k_a and k_b: Check if $IBS[x, k_a]$ and $IBS[y, k_b]$ are marked for all base stations x which are neighbors of station a, and for all base stations y which are neighbors of station b; If not, reject the call and terminate the procedure.
5. Check adjacent-channel interference between station a and b and all their radio neighbors x and y, respectively: For all base stations x and y, such that $TN[a, x]$ and $TN[b, y]$ are marked, check if $IC[x, k_a\pm1]$ and $IC[y, k_b\pm1]$ are all marked. If not, reject the call and terminate the procedure.
6. Check conflicts on the choice of new channels: If p_a and p_b are the same cluster, or if stations a and b are neighbors, then if k_a and k_b are the same channel take a new pair of channels from the list of channels obtained in Step 3 and should go to Step 4. Otherwise, go to Step 7.
7. Find available labels for the terminals: Scan $TB[a]$ and $TB[b]$ and find labels n_a and n_b, such that $TB[a, n_a]$ and $TB[b, n_b]$ are marked; Assign the labels to terminals t_a and t_b, respectively; If both labels are not available, reject the call and terminate the procedure.
8. Update data structures to account for newly assigned channels: Delete $IC[p_a, k_a]$, $IC[p_b, k_b]$, $IBS[a, k_a]$, $IBS[b, k_b]$, $IAL[a, k_a, n_a]$, $IAL[b, k_b, n_b]$, $IP[f_a, n_a]$, and $IP[f_b, n_b]$; Increment $VCPO[f_a]$, $VCPO[f_b]$, $VCN[a]$, and $VCN[b]$; Put a into the $TP[f_a]$ and $TB[b, n_b]$; Put b into $TP[f_b]$ and $TB[a, n_a]$; Put n_a into $TL[f_b, n_b]$, and n_b into $TL[f_a, n_a]$; Put t_a into $TT[f_a, n_a]$, and t_b into $TT[f_b, n_b]$; Terminate the procedure.

Call Termination:

Assumptions: Terminal t_a on platform f_a wants to terminate a call which has been in progress between terminals t_a and t_b. Before the call termination, terminals t_a and t_b have been using channels k_a and k_b, respectively. Terminal t_a is under control of base station a which belongs to cluster p_a. Terminal t_b is under control of base station b which belongs to cluster p_b. During the call, terminals t_a and t_b have labels n_a and n_b assigned, respectively. The following procedure releases channels and updates the data structures:

1. Mark $IC[p_a, k_a]$, $IC[p_b, k_b]$, $IBS[a, k_a]$, $IBS[b, k_b]$, $IP[f_a, n_a]$, $IP[f_b, n_b]$, $IAL[a, k_a, n_a]$ and $IAL[b, k_b, n_b]$.
2. Delete $TT[f_a, n_a]$, $TT[f_b, n_b]$, $TB[a, n_a]$, $TB[b, n_b]$, $TL[a, n_a]$, and $TL[b, n_b]$.
3. Decrement $VCPO[f_a]$, $VCPO[f_b]$, $VCN[a]$ and $VCN[b]$; Terminate the procedure.

Hand-Off:

Assumptions: Mobile platform f_a comes into the cell covered by base station a from the cell covered by base station c. Some terminals on platform f_a may be involved in calls with other terminals which may belong to different platforms, covered by different base stations. Base stations a and c belong to clusters p_a and p_c, respectively.

1. Check if the capacity of the base station a is reached: If $N_{max} = VCN[a]$, reject the hand-off and terminate the procedure.
2. Find the number of currently active terminals on platform f_a, i.e. the number of necessary hand-offs: Read l_a from $VCPO[f_a]$.
3. Determine all l_a "old" labels n_c of the active terminals on platform f_a, which were assigned to the terminals while platform f_a was within cell c: Scan $IP[f_a, n_c]$ for all unmarked values of n_c, when $1 \le n_c \le N_{max}$.
4. Determine all active terminals t_a on platform f_a: For all labels n_c determined in step 3, find terminals t_a such that $IP[f_c, n_c]$ is unmarked and $TT[f_a, n_c] = t_a$; Put the values t_a into the buffer TERMINALS, and put the corresponding values n_c into the buffer LABELS for further processing. The correspondence between the notations of terminals and their labels must be preserved.
5. Determine new channels within the "new" cell covered by base station a by searching for new channels until l_a or all available channels are found and using each channel only once in a cluster to prevent co-channel interference within the cluster: Scan $IC[p_a, k_a]$ for marked k_a, and $1 \le k_a \le K$; If such k_as do not exist, reject the hand-off and terminate the procedure.
6. Determine channels k_a for use in station a, such that co-channel interference with the adjacent base stations are satisfied: Scan $IBS[x, k_a]$ for all marked channels k_a, for all radio neighbors x of station a. If such channels cannot be found, reject the hand-off and terminate the procedure.
7. Select channels k_a for use in station a so that they satisfy the adjacent-channel interference constraints within cluster p_a of station a: Select channels k_a such that $IC[p_a, k_a \pm 1]$ are marked; If such channels cannot be found, reject the hand-off and terminate the procedure.
8. Check for conflicts on channels in the same cluster: If p_a and p_b are the same cluster, or if stations a and b are radio neighbors, discard all selected channels k_a such that $k_a = k_b$.
9. Assume that there are l'_a available channels in cluster p_a, which have passed all the checks for co-channel and adjacent-channel interference. Thus, $min(l'_a, l_a)$ currently active terminals can be handed-off successfully to cell a. The following activities update the data structures: For $min(l'_a, l_a)$ available "new" channels k_a, delete the $IC[p_a, k_a]$, and mark the $IC[p_c, k_c]$. Delete the $IBS[a, k_a]$ and mark the $IBS[c, k_c]$. Delete the $IP[f_a, n_a]$, and mark the $IP[f_a, n_c]$. Put t_a into the $TT[f_a, n_a]$, and delete the $TT[f_a, n_c]$. Put the contents of the $TB[c, n_c]$ into the $TB[a, n_a]$, and put the contents of the $TL[c, n_c]$ into the $TL[a, n_a]$. Delete the $TB[c, n_c]$ and the $TL[c, n_c]$. Increment the $VCN[a]$ and decrement the $VCN[c]$ by $min(l'_a, l_a)$. Terminate the procedure.

10.If $l'_a < l_a$, the multiple hand-off cannot be successful for all the active terminals on platform f_a. Only l'_a calls will be handed-off. The rest will be rejected. For unsuccessful hand-offs, update the data structures: Determine all the pairs of terminals t_a and t_b on platforms f_a and f_b, respectively, which were involved in calls that must be rejected. Determine their labels n_a and n_b, platforms f_a and f_b, current base stations c and b, corresponding clusters p_c and p_b, and channels k_c and k_b, respectively. For all such terminals, mark the $IC[p_b, k_b]$, the $IC[p_c, k_c]$, the $IBS[b, k_b]$, the $IBS[c, k_c]$, the $IP[f_a, n_c]$, and the $IP[f_b, n_b]$. Delete the $TT[f_b, n_b]$, the $TT[f_a, n_c]$, the $TB[b, n_b]$, the $TB[c, n_c]$, the $TL[c, n_c]$, and the $TL[b, n_b]$. For each base station b, decrement $VCN[b]$ by the number of rejected calls which are related to b.

4. INTEGRATED CHANNEL MANAGER

For efficient bandwidth utilization with good transmission quality, we propose an adaptive channel allocation supported by a special purpose hardware device called an *Integrated Channel Manager* (ICM). The ICM may be connected via its input/output lines to a standard system bus of the switch in a cellular network, as shown in Figure 1. In this section, we shall describe a priority coding mechanism

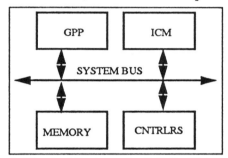

Figure 1: The Architecture of a Cellular Network Switch

suitable for implementation of marking access operations in hardware. This approach provides a speed up of the otherwise time consuming operations. In addition, this approach allows a high degree of parallelism in execution of the channel allocation functions. Later in this section, we shall see how channel allocation can be further improved by introducing a hierarchical organization of the ICM [Ale86, Vuc86, Vuc87, Vuc90]. The priority coding and the hierarchical organization are the major factors which provide a high speed of execution within the ICM.

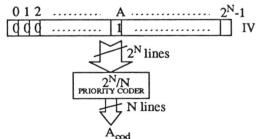

Figure 2.a -Priority Coding of a Main Index Vector

4.1 Priority coding: The algorithms given in Section 3 were described in terms of marking access operations performed on an index vector data structure. Here, we describe an implementation which uses a priority coding mechanism. The priority coding of an index vector is a basic operation of marking access. It determines the address of the first marked IV element in just one access to the index vector. Assume that an IV has 2^N binary elements, and that the first marked element is A, as is shown in Figure 2.a. The $2^N/N$ priority coder is a hardware device which takes the IV as a parallel input and gives an N-bit output A_{cod}, i.e. the encoded address of the marked IV element.

4.2 Hierarchical Organization of Index Vectors: Retrieval of large data bases can be improved additionally by introducing a *hierarchical organization* of the index vector using several levels of auxiliary index vectors [Vuc86, Vuc90]. An example of a marking access based on a two-level hierarchical index vector, is shown in Figure 2.b. Such a vector consists of a main IV and an *auxiliary IV* organized in the following way. Let us assume that the main IV capacity is 2^N elements (bits), and the main IV is stored into 2^{N_1} memory locations. Each of the locations consists of 2^{N_0} bits, where $N = N_0 + N_1$. There is a unique bidirection-

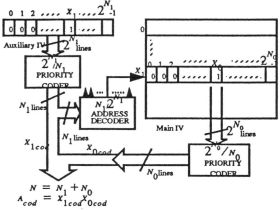

Figure 2.b: Hierarchical Priority Coding of a Two-Level Index Vector

al correspondence between the bit-elements of the auxiliary IV and the main IV locations. If a bit element X_1 of the auxiliary IV contains 0, it means that all 2^{N_0} bits of location X_1 of the main IV contain 0. Therefore, there is no need to scan this location in the search for a marked bit element. If a bit element X_1 of the auxiliary IV contains 1, there is at least one bit element of location X_1 of the main IV which contains 1. Therefore it is necessary to scan this location. Using the auxiliary IV, the number of accesses to the main IV locations is reduced significantly, i.e. only the locations containing marked elements are accessed during scan operation.

In case of hierarchically organized IVs, *priority coding* is proposed to speed up scanning for some data. It determines the location of desired data in one step per level of hierarchy. The total number of hierarchical levels depends on the capacity of the data base which should be scanned for some data.

Let a channel from the given bandwidth corresponds uniquely to an element $IV[A]$ of the main IV, and $A = X_1 \times 2^{N_0} + X_0$, where $A = \{0, 1, ..., 2^N - 1\}$, $X_1 = \{0, 1, ..., 2^{N_1} - 1\}$ and $X_0 = \{0, 1, ..., 2^{N_0} - 1\}$. The information on current channel availability is stored in bit-element X_1 of the auxiliary IV, and in bit-element X_0 of location X_1 of the main IV, as shown in Figure 2.b. In the first step of a search for an available channel, the contents of the auxiliary IV is priority coded. As a result, the code X_{1cod} is obtained. It corresponds to bit-element

X_1 of the auxiliary IV, and represents a higher part of the A_{cod}, the binary coded address of A. In the second step of the search procedure, X_{1cod} is used as an input for the address decoder of the main IV. It selects the corresponding location X_1 of the main IV. This location contains a marked element corresponding to an available channel, and therefore the search for it will be successful. Then, priority coding is applied to the contents of location X_1 of the main IV. As a result, it will give the code X_{0cod}. Concatenation of X_{1cod} and X_{0cod} represents a complete code A_{cod} of the address A.

Further improvement of the efficiency in data base searching can be achieved by introducing more hierarchical layers into the IV structure. If there are Q layers of hierarchy, a data base address A can be represented as:

$$A = X_0 + \sum_{i=1}^{Q-1} X_i \times 2^{N_0 + \dots + N_{i-1}} \tag{1}$$

A restriction $N_i = N'$ may be introduced if lengths of memory blocks used to store auxiliary and main IVs are standardized and mutually equal. Then,

$$A = \sum_{i=0}^{Q-1} X_i \times 2^{i \times N'} \tag{2}$$

The optimal efficiency can be achieved if the number of hierarchical levels is $Q = N/N'$ [Vuc86, Vuc87, Vuc90]. In that case, any marked element in the IV can be found in only Q accesses to the IV structures.

4.3 The Architecture of the ICM: The ICM receives and executes commands from the *General Purpose Processor* (GPP) as any other integrated controller. Upon execution on demand, the ICM returns back to the GPP a status or result of the command. The ICM uses the GND and V lines for standard power supply.

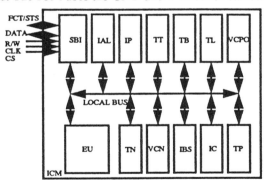

Figure 3: The Architecture of the ICM

The architecture of the ICM, shown in Figure 3, consists of three major functional blocks: *System Bus Interface* (SBI), *Execution Unit* (EU), and *Data Structures*. The SBI is used to control flow of information between the GPP and the ICM. The SBI consists of synchronization and control circuits, and several data buffers for temporary data storage. The GPP addresses the ICM as any other controller on the system bus. The ICM has a unique address. When this address is on the system bus, the ICM is activated via the ICM *Chip Select* (CS) input line. The ICM receives external synchronization clock signals via the CLK line. The GPP controls the flow of information to and from the ICM using the W/R input line. It notifies the ICM if the GPP reads a status from, or writes a command into the ICM internal registers. The ICM also has bidirectional DATA lines used by the

GPP to send parameters to, or receive results of execution from the ICM. Data lines of the system bus are bidirectional and divided into two parts: *Function Lines* (FCT) in the input mode, and as *Status Lines* (STS) in the output mode. The FCT lines are used by the GPP to write a code of the ICM function into the ICM. The STS lines are used by the GPP to get the status of the ICM or the result of the previous ICM execution. A part of the DATA lines is used for transfer of parameters.

The EU is used to execute the adaptive channel allocation functions. These functions are defined by the FCT code received from the GPP. The EU includes integrated scanner devices associated with the main and auxiliary index vectors. The scanners are hierarchically organized so that each scanner corresponds to one and only one *IV*. The main operation within a scanner is priority coding. It is used to speed up searching for an available channel within the *IC* and *IBS*, searching for an available label in the *IAL*, and searching for all the radio neighbors of a base station within the *TN* data structure.

The Data Structures have been described in section 3.2. The *IC*, the *IBS*, the *IAL*, the *IP*, and the *TN* are data structures organized as index vectors. Therefore, we may apply the priority coding mechanism to expedite the handling of these data structures.

5. PERFORMANCE ANALYSIS

The ICM's performance is evaluated in terms of throughput and efficiency of channel allocation. To illustrate the ICM performance, we simulated a network consisting of $S=21$ base stations divided into $P=3$ clusters. The maximal number of available channels (K) has been varied from 1 to 400. The maximal number of wireless terminals (users) supported by a base station (N_{max}) has been varied from 4 to 100. The maximal number of users per mobile platform (L) has been varied from 1 to 10. The maximal number of platforms which can be supported by the network switch (F) is equal to 210. In the initial state, there are no calls in progress, and there are 10 mobile platforms in each cell. The average call arrival rate is 1800 calls/hour. There is one request for hand-off every 400 seconds on the average. The average duration of a call is 3 minutes. The average processing times are: 2μsec for a call termination, 6μsec for a call initialization, and 12μsec for a hand-off. The processing times are estimated based on the number of steps and the duration of channel allocation functions executed by the ICM with a high degree of parallelism.

5.1 Analysis of Throughput: Throughput of the network (T) is a function of different parameters: the offered load (G), the maximal number of available channels (K), the maximal number of users per base station (N_{max}), the maximal number of users per platform (L), the maximal number of platforms in network (F), etc.

In our analysis, throughput is defined as a ratio between *Carried* (G') and *Offered Load* (G). The offered load is a total number of calls which should be supported by the network. It may be defined by the following relation:

$$G = \sum_{i=1}^{S} \sum_{j=1}^{f_i} l_{ij} \tag{3}$$

where l_{ij} denotes the number of calls originating at platform j which is currently within the area of cell i. f_i denotes the number of active platforms in cell i. A platform is active if at least one user on the platform is active.

The total number of active platforms in the network is:

$$F_o = \sum_{i=1}^{S} f_i \qquad (4)$$

However, the network can support at most F platforms simultaneously. It holds:

$$F = \sum_{i=1}^{S} f'_i \qquad (5)$$

where f'_i is the number of platforms in cell i carrying at least one active user.

The following relation shows that a platform which previously did not have any active users can be rejected. If a platform has not already been active when a potential user on the platform wants to make a call, the call will be rejected if the current total number of active platforms in the network is equal to F.

$$f'_i = \begin{cases} f_i & \text{if } \sum_{k=1}^{S} f_k \le F \\ f^0_i & \text{if } f^0_i + \sum_{\substack{k=1 \\ k \ne i}}^{S} f_k = F \end{cases} \qquad (6)$$

where f^0_i denotes the number of platforms which have already been active in cell i. f_k denotes the current numbers of active platforms in all other cells of the network when a new potential user requests service.

Carried load G' represents a total number of successful calls in the network. It can be calculated from the following relation:

$$G' = \sum_{i=1}^{S} \sum_{j=1}^{f'_i} l'_{ij} \qquad (7)$$

where l'_{ij} denotes the number of currently active users on platform j in cell i. A potential user will be rejected if there has already been L users currently active on its platform. Thus, $l'_{ij} = \min(L, l_{ij})$.

We define the throughput of the network as $T = G'/G$. This is a complex function of the listed parameters as well as of the statistics of call durations, the frequency of hand-offs, the sizes of cells, mutual interests for communications between users, etc. Here, we shall estimate a theoretical maximum of throughput subject to three parameters:

1. Maximum number of available channels with frequency reuse, PK;
2. Maximum number of users, limited by the capacity of base stations, SN_{max};
3. Maximum number of users, limited by the capabilities of platforms, FL.

According to these restrictions, the following cases can be discussed:

1. If $PK \le SN_{max} \le FL$, or $PK \le FL \le SN_{max}$, the dominant restrictive factor is a limited number of channels. In such a case, the throughput has a theoretical maximum equal to 100% for offered loads $G \le (PK)/2$. For higher loads, theoretical maximum of the throughput is equal to $T_{max} = PK/(2G)$.
2. If $SN_{max} \le PK \le FL$, or $SN_{max} \le FL \le PK$, the dominant restrictive factor is a limited capability of base stations to support wireless terminals. The throughput has a theoretical maximum equal to 100% for loads $G \le (SN_{max})/2$. For higher loads, theoretical maximum of the throughput is equal to $T_{max} = SN_{max}/(2G)$.
3. If $FL \le SN_{max} \le PK$, or $FL \le PK \le SN_{max}$, the dominant restrictive factor is a

limited capability of the switch to support mobile platforms, and a limited capability of platforms to support multiple users. A theoretical maximum of the throughput is equal to 100% for loads $G \leq (FL)/2$. For higher loads, the throughput has a maximum given by the relation: $T_{max} = FL/(2G)$.

Thus, an upper bound to throughput can be expressed as

$$T_{max} = \min(1, \frac{FL}{2G}, \frac{SN_{max}}{2G}, \frac{PK}{2G}) \qquad (8)$$

Figure 4 represents the behavior of the throughput (T) vs. maximal number of available channels (K) and maximal number of users per base station (N_{max}), when the maximal number of users per platform L is equal to 10.

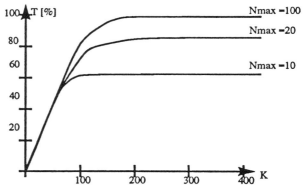

Figure 4: Total Throughput (T) vs. Number of Channels (K) and Maximal Number of Users per Base Station (N_{max}) when L= 10

It can be seen that the throughput increases when N_{max} increases, i.e. when base stations can support more users. The diagram referring to the case when $N_{max}=100$ represents an example when PK is a dominant restriction according to (8). The diagrams referring to the cases when N_{max} is equal to 10 and 20, represent examples of a limited throughput due to the combined influence of PK and SN_{max} as dominant restrictive factors. For low values of K, PK is a dominant restrictive factor of throughput in (8). Thus, a higher throughput can be achieved by increasing K. However, when $K \geq (SN_{max})/P$, N_{max} becomes a dominant restrictive factor. No matter how many additional channels are available, the calls are rejected if the current number of active calls is $G' = SN_{max}/2$. In our example, it occurs with $K \geq 21 \times 20/3 = 140$ channels when $N_{max} = 20$, or with $K \geq 21 \times 10/3 = 70$ channels when $N_{max} = 10$.

Figures 5.a and 5.b depict the influence of the third restrictive factor (FL) on the throughput (T). In Figure 5.a, N_{max} is relatively high and it is not a dominant restriction to the throughput. When $L=10$, PK is the only dominant restriction. For low values of K, the throughput increases with K. When $K \geq (2G)/P$, $T=100\%$. For lower values of L, there is a combined influence of PK and FL on the throughput. When $K \leq (FL)/P$, a dominant restriction to the throughput is PK. Thus, a higher throughput can be achieved by increasing K. When $K > (FL)/P$, FL becomes a dominant restriction to the throughput. In our examples shown in Figure 5.a, it occurs with $K \geq 210 \times 4/3 = 280$ channels and $L=4$, and when $K \geq 210 \times 1/3 = 70$ channels and $L=1$. Any further increase of K will not increase the throughput. If $G' = (FL)/2$, every new call is rejected.

In Figure 5.b, all three restrictive factors can be seen. N_{max} is relatively low and it becomes a dominant restriction of throughput when $K \geq (SN_{max})/P$ and $L>1$.

In our example, it happens when $K \geq 21 \times 10/3 = 70$ channels. When $L=1$ and $K \geq (FL)/P$, factor FL becomes a dominant restriction of the throughput. In the example, it occurs when $K \geq 210 \times 1/3 = 70$ channels. For $K<70$ channels, PK is a dominant restrictive factor of the throughput.

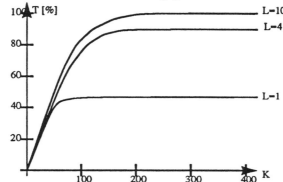

Figure 5.a: Total Throughput (T) vs. Number of Channels (K) and Maximal Number of Users per Platform (L) when $N_{max} = 100$

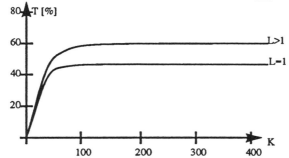

Figure 5.b: Total Throughput (T) vs. Number of Channels (K) and Maximal Number of Users per Platform (L) when $N_{max} = 10$

From this analysis, we can conclude that all three restrictive factors should be adjusted simultaneously to provide the maximal throughput. After the influence of various network parameters is evaluated, the dimensions of the ICM data structures may be determined.

5.2 Analysis of Efficiency: The efficiency of call initiation and hand-off has a significant influence on transmission quality and overall throughput. Therefore, it represents an important objective in the design of cellular networks. In this analysis, the efficiency will be expressed as the number of accesses to the data structures containing information on current channel availability, during the execution of the adaptive channel allocation functions. To show the efficiency improvement achieved by implementation of the ICM, we compare the obtained results with the corresponding ones which are derived by a similar algorithm implemented in software. The latter algorithm is based on sequential searching of the same data structures for available channels. The comparison is expressed in terms of a ratio between the numbers of accesses to the data structures obtained by the ICM and by sequential searching. The ratios are calculated for the numbers of accesses per call initialization (R_c) and per hand-off (R_h).

In Figures 6.a and 6.b, families of curves represent values R_c and R_h when K and

256

L are varied. For both R_c and R_h, the ICM advantage in efficiency can be seen. With sequential searching, the average number of accesses to a data structure is equal to $M/2$, where M denotes the capacity of the data structure. With the ICM, the number of accesses to a data structure is always equal to Q, where *Q* denotes the number of the ICM hierarchical levels. The ICM efficiency is considerably higher than the efficiency of the sequential searching since $Q \ll M/2$.

For higher values of *L*, the superiority of the ICM is even more apparent. It is a consequence of the fast retrieval of the ICM data structures. For *L*>1, the data structures are more complex and have higher capacities than in the case when *L*=1. It deteriorates significantly the efficiency of a sequential retrieval, while the reduction of ICM efficiency may be neglected. The ICM efficiency is reduced only if an additional level of hierarchy is introduced into the data structures. It increases the number of accesses to each of the data structures by 1.

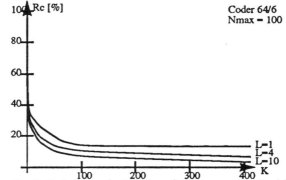

Figure 6.a: Ratio between the Number of Accesses per Call with the ICM and Sequential Searching vs. K and L

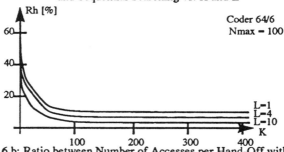

Figure 6.b: Ratio between Number of Accesses per Hand-Off with the ICM and Sequential Searching vs. K and L

In Figures 7.a and 7.b, the families of curves represent ratios R_c and R_h when *K* and type of priority coders are varied. Both R_c and R_h are descending functions of *K*. As the number of channels increases, the ICM shows its advantage over the sequential searching. The ICM finds an available channel (if it exists) in only a few accesses to the data structures per call, and the number of the accesses is constant. It does not increase when *K* is increased. In sequential searching, when *K* is increased, the probability of successful calls is increased but the search for an available channel takes longer. Thus, the efficiency is reduced.

In Figure 7, there is a break-point after which no additional channels can improve the efficiency of the ICM. This break-point is due to the other two factors (*FL* and SN_{max}). The calls are rejected because base stations or the switch cannot support them even if there are available channels. If N_{max} and/or L are higher, the break-point appears at the higher values of K. It implies that more calls can be supported by base stations. This estimate is very useful in a network design and evaluation of a trade-off between the limits of the bandwidth, base station and the switch capacities subject to the predicted offered load.

In Figures 7.a and 7.b, we see that the type of priority coders within the ICM also makes an influence on the values of R_c and R_h. If a higher order coding (e.g. the 64/6 coding, compared to the 32/5 or the 16/4 coding) is implemented, there are fewer levels of hierarchy in the ICM. The number of auxiliary IV s is reduced. Therefore, the number of accesses to data structures is reduced, and the ICM is more efficient. However, the implementation of high order priority coders may be impractical and even infeasible. Generally, a trade-off between the type of coders and the number of hierarchical levels in the ICM should be made.

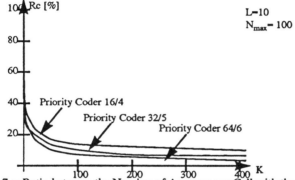

Figure 7.a: Ratio between the Number of Accesses per Call with the ICM and Sequential Searching vs. Number of Channels (K) and Type of Coder

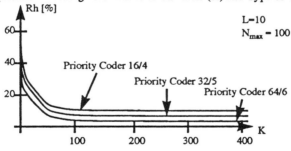

Figure 7.b: Ratio between Number of Accesses per Hand-Off with the ICM and Sequential Searching vs. Number of Channels (K) and Type of Coder

6. CONCLUSION

In this paper we propose a new concept for an adaptive channel allocation in cellular networks. The concept is suitable for implementation in hardware as a special-purpose VLSI component called an Integrated Channel Manager (ICM). It can be connected to the system bus of the cellular network switch similarly to any other integrated controller. The ICM executes the following channel allocation functions: Reset, System Initialization, Registration, Call Initialization, Call Termination, and Hand-Off.

The ICM is an efficient hardware device with a high degree of parallelism. This hardware is based on a logical structure called Index Vector and a memory access mechanism called Marking Access. It is used to speed up the search for available channels which satisfy co-channel and adjacent-channel interference constraints during adaptive channel allocation. Also, it provides a fast call rejection when the switch cannot support the call. This capability is advantageous, especially in case of a heavily loaded network when the processing time of the switch should not be wasted on calls which will be rejected anyway due to unavailability of channels. Further improvement of the ICM efficiency can be achieved by introducing a hierarchical organization into the index-vector-based data structures and the associated execution unit.

The ICM component can be easily adapted to interface different system buses. The ICM modularity also enables its easy adaptation to different radio bandwidths and other network parameters. The extension of the ICM instruction set is also possible.

The ICM's application does not depend on any specific property of a radio channel. The channels are modeled as a shared resource subject to a set of allocation constraints. Therefore, the ICM's application is not limited only to cellular networks with the frequency division multiple access (FDMA). It is also suitable and recommended for implementation in the code division multiple access (CDMA) concept. In that case, instead of channel allocation, the ICM allocates spreading codes which denote sequences for orthogonal spread-spectrum transmissions [Par89].

REFERENCES

[Ale86] T. Aleksic, "A Relational Data Base Computer," Internal Documentation, Department of EE/CS, University of Beograd, Yugoslavia, 1986.

[Eri88] H. Eriksson, "Capacity Improvement by Adaptive Channel Allocation," Proc. of IEEE Globecom-88, November 1988.

[Goo90] D. J. Goodman, "Cellular Packet Communications," IEEE Trans. on Communications, Vol. 38, No. 8, August 1990.

[Nan90] S. Nanda, and D. J. Goodman, "A Paradigm for Distributed Channel Allocation," Proc. of the Second Rutgers WINLAB Workshop on Third-Generation Wireless Information Networks, East Brunswick, NJ, October 1990.

[Nan91] S. Nanda, and D. J. Goodman, "Dynamic Resource Acquisition: Distributed Carrier Allocation for TDMA Cellular Systems," Proc. of the IEEE GLOBE-COM-91, Phoenix, AZ, December 1991.

[Par89] J. D. Parsons, and J. G. Gardiner, *Mobile Communications Systems*, Blackie, Halsted Press, 1989.

[Rap90] S. Rappaport, "Models for Call Hand-Off Schemes in Cellular Communication Network," Proc. of the Second Rutgers WINLAB Workshop on Third-Generation Wireless Information Networks, East Brunswick, NJ, October 1990.

[Vuc86] J. Vucetic, *RAMs with Marking Access*, Master's Dissertation, Department of EE/CS, University of Beograd, Yugoslavia, April 1986.

[Vuc90] J. Vucetic, "RAMs with Marking Access," Supercomputing-90, Poster Presentation, New York, NY, November 1990.

[Vuc91] J. Vucetic, "Performance Analysis of Index-Vector-Based Algorithms for Fast Adaptive Channel Allocation in Cellular Networks," Proceedings of the IEEE GLOBECOM-91 Conference, Phoenix, AZ, December 1991.

[Vuc92a] J. Vucetic, and D. Dimitrijevic, "Extended Integrated Channel Manager (EICM) - An Architecture for Fast Adaptive Channel Allocation in Cellular Networks with Multi-Terminal Platforms," Proc. of the IEEE ICC-92 Conference, Chicago, IL, June 1992.

[Vuc92b] J. Vucetic, and D. Dimitrijevic, "Adaptive Channel Allocation in Cellular Networks with Multi-User Mobile Platforms," Proc. of the Third WINLAB Workshop on 3rd Generation Wireless Networks, East Brunswick, NJ, April 1992.

Asymptotic Bounds on the Performance of a Class of Dynamic Channel Assignment Algorithms

Jens Zander
Radio Communication Systems
Royal Institute of Technology
ELECTRUM 207
S- 164 40 STOCKHOLM-KISTA, SWEDEN

and

Håkan Eriksson
R&D Dept
Ericsson Radio Systems AB
S- 164 80 STOCKHOLM-KISTA, SWEDEN

Abstract

Dynamic channel assignment adapting both to traffic variations as well as to changing mobile locations is investigated. The study focuses on the asymptotic performance, i.e. the performance in systems with large traffic loads. Both upper and lower bounds for the asymptotic performance of optimum algorithms within the class of reuse-type Dynamic Channel Assignment (DCA) are found for a simple propagation model. As performance measure, the probability of assigment failure (intra-cell handoff failure) is used. Results show that the capacity we may expect to achieve with these algorithms in the asymptotic case is just above twice of the capacity of a fixed channel allocation scheme.

1. Introduction

In modern cellular radio systems, the number of subscribers is expected to grow rapidly. Increasing the capacity of these systems, i.e. the number of subscribers per area (or volume) unit that can be handled at some predefined level of service quality, is of paramount importance. The most straight-forward method to increase the capacity of the system within a given bandwidth, is to reduce the cell size. The main drawback of this solution is the large number of base stations required. This number increases roughly proportional to the achieved capacity. An alternative solution is to reuse the available spectrum more efficiently. This means that the same channel, time-slot or code should be spatially reused more often. A more dense spectrum reuse, in turn, increases the level of intereference. Thus, either the transmission scheme has to be more tolerant to co-channel interference, or the co-channel interference has to be reduced by efficient spectrum planning schemes. The latter method will be the topic of this paper.

Traditional cell planning[1], where channels are allocated to mobiles and base stations according to predictions about propagation and traffic load conditions have been shown to be reasonably effective in systems with high and constant loads. These fixed plans, however, fail to provide high capacities when faced with changing traffic and propagation conditions (due to mobile movements). These problems have brought *Dynamic Channel Assignment* (DCA) schemes into the focus of attention of the cellular radio community. A DCA-system will, instead of solely relying on a-priori information, use knowledge about actual conditions.

A classification of different DCA-algorithms has been proposed by Beck and Panzer[2]. They subdivide the adaptability of DCA-schemes into three cathegories, adaptability to *traffic, channel reusability* and *interference* . In the literature, so far most interest has been devoted to traffic adaptation. Such schemes rely on propagation models or signal level measurements to determine which *pairs of cells* that are compatible in the sense that the same channel may be used simultaneously in that pair. This pair-wise compatibility information may be represented by a matrix, the *compatibility matrix*. In [3], the maximum packing (MP) concept is proposed which finds the minimum number of channels that are required to handle given number of calls. In [5-6] the asymptotic behaviour of traffic adaptive DCA, i.e. the performance as the number of calls and channels approaches infinity, has been investigated. Results show that for a traffic load stationary in time (but not necessarily spatially uniform), the capacity gain derived from dynamic channel assignment becomes less and less as the number of channels increases. Optimum traffic adaptive DCA schemes thus have the same asymptotic performance as fixed channel allocation.

It is obvious that the compatibility matrix gives a rather coarse description of the interference situation since it has to be based on "worst case" assumptions about mobile locations and propagation conditions. If two cells are declared to be compatible by means of the compatibilty matrix, no combinations of shadowing and unfavorable mobile positions may be allowed to cause unacceptable interference in either cell. To achieve this, certainly a considerable margin in the design signal-to-interference ratio will be required. In turn, this may cause a large penalty in capacity. Algorithms that take actual propagation (and interference) conditions into account are represented by the vertical, "interference", direction in fig 1. These interference variations are mainly caused by changing propagation conditions(fading, shadowing) and by varying traffic load in co-channel cells. The problem of finding an explicit optimum DCA-scheme fully exploiting this knowledge has proven to be a difficult one. Very little general results are known and a compact description of the optimum "INFOPT" algorithm, i.e the algorithm that maximizes the number of served customers at a given quality level is not known. However, several explicit methods have been proposed and results are promising.

Since the instantaneous interference situation is difficult to assess, we will be limited to use DCA-schemes that rely on more or less rough estimates of the interference level. The compatibility matrix discussed above represents one, rather coarse and pessimistic method to achieve this. Simple ways to improve our estimate of the interference situation have been proposed. It is easy to understand that a channel that is used by mobiles close to their respective base stations in general have a higher received signal level than more distant mobiles and are thus able to tolerate a higher interference level. These channels may therefore be "packed" closer than channels used by mobiles located close to the cell boundary. One could exploit this fact by extending the concept of cell compatibility. Let r_{ij} denote the distance between mobile j in cell i and its base station. A modified (and unfortunately more complex) compatibility description would tell us if mobiles with distances r_{ij} and r_{kl} may use the same channel in cell i and k. DCA schemes adapting to these conditions constitute the class of *reuse-type* algorithms. The most straigth-forward application of this concept is to divide the cell into zones, each with a different reuse pattern. This method has been coined *reuse-partitioning* [4,8].

In this paper, we will elaborate on this concept and turn our attention to reuse-type DCA-algorithms that also exploit traffic adaption. The optimum scheme in this class is denoted the "Reuse-MP" algorithm[2]. To investigate the asymptotic performance of this strategy for a simple propagation model is the purpose of this paper. Since an efficient description of this strategy is not known, we derive both an upper and a lower performance bounds. The bounds are then compared with the asymptotic performance of fixed channel assignment.

2. Models and performance measures

Throughout the paper, we will assume that we have a cellular radio systems with a large number of base stations located on a hexagonal grid. The mobiles roam in the entire service area and are assumed to be uniformly distributed over this area. The average received signal strength is assumed to decrease monotoneously with the distance between base and mobile stations according to

$$P_r = \frac{P_0}{r^a} \tag{1}$$

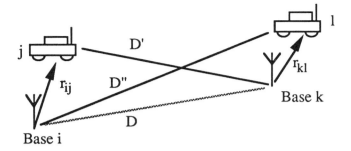

Fig 1. Geometrical considerations of channel (frequency) reuse

The mobiles are assumed to communicate with the closest base station, i.e the base station with the highest received signal level. The *cell* served by a base station is simply the collection of points closer to that base station than any other base station. In our case the cells are hexagons with a base station at the center of each cell. We will in the following make the assumption that omnidirectional antennas are used. Without loss of generality we will further assume that the side (or radius) of the cells is unity.

Within our system we assume to have M independent(orthogonal) communication channels available. This means that the system design is assumed to be such that adjacent channel interference may be neglected. For the sake of simplicity we confine our analysis to the down-link (base-to-mobile) channel. The channels are geographically reused in order to a achieve a high capacity. The frequency reuse is limited only by the growing co-channel interference in the system. Let us assume that the j:th mobile is communicating with base station i at a distance r_{ij} from its base station (fig 1). Suppose we want to reuse the same channel at some distant base station k to serve a mobile l. In order to keep the signal to interference ratio (C/I) in both links above some prescribed threshold γ_0, the geographical distance between the cells has to be large enough. Let us first considering the (down) link between base station i and its mobile j. The reuse distance requirement for this link may be expressed as

$$D \geq \eta*(\gamma_0)\, r_{ij}, \qquad (2)$$

where the constant $\eta*$ is chosen to allow satisfactory operation also in "worst case" situation regarding propagation (fade margin), mobile direction, and interference conditions. Relation (2) may be interpreted as a pair-wise compatibility requirement for our system. A channel may be used in a cell k, if this cell is located at a distance D fullfilling

$$D \geq \eta* r_{ij} = D_{ij} \qquad (3)$$

where D_{ij} is denoted the *reuse distance* associated with mobile ij. The approximation $D' \approx D$ obviously works well for large reuse distances. For small D:s, the selection of $\eta*$ may in some cases have to be very conservative, such that (3) yields pessimistic results.

In order to relate our results to existing systems, we will use a reference system using a conventional, repeatable, hexagonal cell plan with N channel groups. The M channels are assumed to be equally divided among these groups, resulting in an assignment of M/N channels per group. N is usually referred to as the *cluster size* of the cell plan. Such a cell plan is designed to achieve an adequate C/I-level in all mobile locations, including mobiles at the cell border (r = 1). In the following we will therefore assume that the C/I-requirements and the propagation constant a are exactly such that we may (just barely) use a fixed cell plan of cluster size N. By "just barely" we mean that increasing the C/I-requirement to $\gamma_0 + \varepsilon$ for any positive ε, would require a cell plan with larger cluster size than N. N will be a compact way to decribe the combined effects of C/I-requirements and the propagation conditions.

The problem of dynamic channel assignment is that of assigning a channel to each active mobile such that the constraint (3) is satisfied on every link. The traditional performance analysis for cellular radio systems assumes Poisson call arrivals and exponential call durations. Further, the cell size is assumed to be that large, that mobiles are usually assumed to move only neglectable distances during a call. The system is then modelled as a Markovian queueing system with blocking. The main performance measure in the traditional analysis is the *blocking probability,* i.e. the probability that a newly arriving call is not assigned a channel[12]. In a small cell environment, the distance traversed during a call may not be neglectable compared to the dimensions of a cell. The C/I-ratio may change considerably and utilizing a dynamic channel assignment scheme may require many channel reassigments. The success of these reassignments, (*intra-cell handoffs*) will be of vital importance to the quality of service provided by the system.

In order to determine the efficiency of these reassignments, we will study a system in operation at some given instant of time and determine how well a dynamic channel assignment algorithm may accomodate the calls beeing in progress. Let us assume that the *active* mobiles, i.e. those with calls in progress, are uniformly distributed in each cell. The number of calls in progress in cell i is denoted U_i. The U_i:s are all assumed to be independent Poisson distributed variables with expectation 1 (calls per cell). We also define the *relative traffic load*, ρ, defined as

$$\rho = \frac{\lambda}{M} \tag{4}$$

Now, for some given assignment scheme and some arbitrary instant of time, let the random variable Z_i denote the number of calls that are not assigned a channel in cell i. We will assume that the assignment schemes used are "fair" in the sense that the resulting Z_i :s have identical statistial properties in all cells. Our measure of performance will be the *assignment failure rate* defined as

$$v = \frac{1}{\lambda} E[Z], \tag{5}$$

which is the "relative" number of calls that the DCA-algorithm fails to assign a channel at this given instant of time. For large loads λ, we may show that for some randomly chosen mobile the quantity v approaches v^*, the probability that some given call is not assigned a channel. Since asymptotic results are the main concern of this paper we will refer to v^* as the *asymptotic probability of assignment failure*.

3. A lower bound on the probability of assignment failure

Since no explicit form of the Reuse-MP algorithm is known, we will in this section derive a upper performance bound for this algorithm and thus all other algorithms of this class. We state our result as

Proposition 1:

For any reuse-type dynamic channel assignment algorithm, the asymptotic probability of assignment failure in a planar cellular network is lower bounded by the following expression:

$$
v^* = \lim_{M\to\infty} v(\rho) \geq
\begin{cases}
0 & \rho \leq \rho^* \\[2mm]
\dfrac{2\sqrt{3}}{\pi N} \dfrac{(\rho-\rho^*)}{\rho\rho^*} & \rho > \rho^*,
\end{cases}
$$

where

$$
\rho^* = \frac{24\sqrt{3}}{5\pi N\left(1+\dfrac{4\pi}{5\sqrt{3}N^2}\right)} \approx \frac{2.65}{N\left(1+\dfrac{1.5}{N^2}\right)}
$$

is called the "cut-off load". N denotes the number of channel groups (cluster size) required by a reference system using fixed assignment and a symmetric, repeatable, hexagonal cell plan.

◊

A proof of this proposition is given in [9].Note that the assumption that the active mobiles are Poisson distributed is never used explicitly thanks to the central limit theorem. In the first case ($\rho\leq\rho^*$) the lower bound is of course trivial since n has to be a positive number. This is due to the fact that the sum B can be confined to be arbitrary "close" to its average m_B just by increasing the number of channels. The bound for n can thus, in turn, be made arbitrary small by just increasing the number of channels. For loads above the "cut-off" load ρ^*, however, the theorem tells us that n never can be smaller than some positive number, regardless of how much we increase the number of channels.

It is easy to see that for a fixed cell design using N channel groups, the load ρ is limited to 1/N. Below this load, we may reduce n to an arbitrary small value for a fixed channel assignment simply by increasing M. Above this load, icnreasing M will do no good. We note that the "cut-off" load, ρ^* for moderate to high N is very close to 2.6/N, thus exceeding the upper load limit of a fixed allocation scheme by a factor of 2.6.

4. An upper bound on the achievable probability of assignment failure

The technique used in lower bounding the performance of reuse-DCA:s in the previous chapter, used continous reuse distances. Obviously, it is in general not possible to implement these distances in a discrete cell plan. In order to get an upper bound on the asymptotic performance of the Reuse-MP algorithm, we propose a specific construction ("generalized" reuse partitioning) using hexagonal cell patterns. We have

Proposition 2:

There exist reuse-type dynamic channel assignment algorithms for planar cellular networks for which the asymptotic probability of assignment failure is not more than :

$$
v^* = \lim_{M \to \infty} v(\rho) \leq
\begin{cases}
0 & \rho < \rho_0 \\
1 - \dfrac{\rho_0}{\rho} & \rho \geq \rho_0,
\end{cases}
$$

where the "critical load" ρ_0 is given by

$$\rho_0 = \frac{N}{S(N)} = \frac{N}{\displaystyle\sum_{i=1}^{K} n_i(n_i - n_{i-1})}$$

where N again denotes the number of channel groups required in the reference system with fixed channel assignment and a symmetric hexagonal cell plan. The n_i are the possible cluster sizes in such a cell plan (eq 4b) ordered such that

$$n_1 < n_2 < ... \, n_K$$

where we have defined $n_K = N$ and $n_0 = 0$.

The proof of this proposition is based on a generalized reuse-partitioning scheme using a symmetric hexagonal reuse pattern. The cells are divided into K zones with radii $r_1, r_2, ...$ 1 (fig 2). With zone i we associate a fixed hexagonal channel plan with cluster sizes $n_i = 1,...$ N. Each zone i, or channel plan i is allocated a number of channels, c_i, where the sum of the c_i adds up to M, the total number of available channels.

A proof of this proposition is given in [9]. The problem of chosing the proper channel allocations i.e. the number of channels that are to be assigned to each zone has be discussed in some detail in [8]. Allocating channels proportional to the expected number of call in each zone is, as we see from the proof of the proposition, the optimum choice for the asymptotic case.

Using the same technique way may now easily state the following simple corollary:

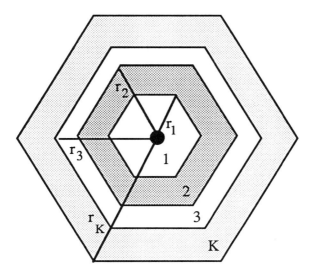

Fig 2. "Generalized" reuse partitioning in hexagonal cells[8].

Corrollary 2:

The asymptotic probability of no channel assignment in the reference system using fixed assignment is bounded by :

$$v^* = \lim_{M \to \infty} v(\rho) \leq \begin{cases} 0 & \rho < \rho_{fix} \\ \\ 1 - \dfrac{\rho_{fix}}{\rho} & \rho \geq \rho_{fix}, \end{cases}$$

where

$$\rho_{fix} = \frac{1}{N},$$

and N is the reference cluster size defined as in the previous propositions.

The results from proposition 1 and 2 and corrollary 2 are summarized in fig. 3 and 4. In fig. 3 we illustrate the upper bound (Prop 2), the lower bound (Prop 1) and the asymptotic performance of the fixed assigment reference system. Clearly the asymptoptical performance of the optimum ("Reuse-MP") algorithm has to be confined to the shaded area between the upper and lower bounds. Fig. 4 shows the behavior of the "cut-off" load ρ^* and the critical loads ρ_0 and ρ_{fix} as function of the cluster size N. The diagram illustrates the that the two quotients ρ^*/ρ_{fix} and ρ_0/ρ_{fix} approach the values 2.6 and 2 respectively as N increases.

Fig. 2 displays a relatively large difference between the upper and lower bounds. The main reason for this is that the bounding technique used in proposition 1, essentially allows "continous" reuse distances, whereas the we in the lower bound are confined to a particular cell layout and a discrete reuse pattern.

We belive that the constructive lower bound may be improved upon. An algorithm allowing for irregular patterns could achieve better results, in particular for low and moderate values of N. In addition, the reuse distance for each zone is chosen such that C/I requirements are met in the corners of the hexagonal zones in fig 1. Only very few mobiles have distances close to the corner distance. Therefore, we cannot exclude the possibility that other (smother) zone shapes could provide better results.

5. Discussion

In the paper we have derived both upper and lower bounds for the asymptotic probabilty of assignment failure in cellular radio systems. The critical loads in the upper and lower bounds were shown to be rather close for large C/I requirements. As we compare the critical loads in the lower and upper bounds to critical load of fixed assignment, a capacity gain of at least 2, but at most 2.6, is feasible by using a reuse-MP scheme in a large cellular system. This may seem as a rather moderate gain, compared to some of the results found in the evaluation of systems with a finite number of channels.

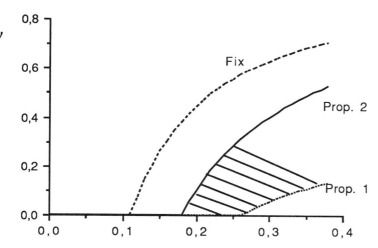

Fig 3. Upper and lower bounds for the asymptotic probability of no channel assignment v* for the optimum reuse-MP algorithm compared with the the asymptotic probability of no channel assignment for fixed channel assignment(N=9).

One should, however, keep in mind the nature of these asymptotic results. As already shown in [5,6], the relative gain by using dynamic channel assignment drops as the number of available channels (and traffic load) increases. This is a straigth-forward consequence of the law of large numbers, which tells us that for a constant ρ, the variance of the number of calls divided by the number of channels will go to zero as we increase the number of channels. As a result the number of calls will be very close to the expected number of calls, a relatively very small number of channels are needed to cope with load variations. The gain by DCA is in consequence diminishing.

As we observe in the proofs of proposition 1 and 2, the same holds also for Reuse-MP type algorithms. The gain achieved by using a DCA-algorithm adaptive to traffic variations diminishes and the remaining gain has to be attributed to the exploitation of the channel reusability. The results in the paper are thus not in contradiction with

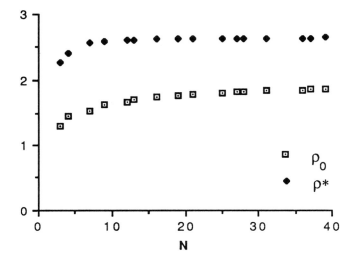

Fig 4. Comparision between the relative "cut-off" load $\rho*/\rho_{fix}$ and the relative critical load ρ_0/ρ_{fix} as function of cluster size N.

previous results on DCA. In fact, DCA should achieve best results when used in systems with a low to moderate number of channels. Here relative traffic variations are considerable, and fixed systems have to be designed with extensive capacity margins to achieve acceptable assignment failure performance. This additional effect enable DCA schemes to provide for larger capacity gains compared to fixed schemes [7].

Reuse-type DCA-schemes still depend on the concept of pair-wise mobile compatibility. In the selection of $\eta*$ in section 2, we require considerable margins regarding fading and traffic conditions in other cells(i.e. the number of interferers). Considerable performance improve-ments compared to the systems described here are therefore still to be expected if information about actual interference conditions can be exploited.

Throughout the paper omnidirectional antennas are used. We believe, however that this assumption is not critical and similar results may be derived for sectorized systems. Also, we have found that the assumption that the number active mobiles in a cell is Poisson is not critical. This assumption is actually used only to construct a simpler proof in proposition 2.

6. Acknowledgement

The authors would like to thank Magnus Frodigh for all the discussions, proof reading and constructive critisism regarding these results.

7. References

[1]   Lee,W.C.Y, "Elements of Cellular Radio Systems Design", *IEEE Trans Veh Tech,* VT-35, No 2, 1985.

[2]   Beck, R., Panzer, H., "Strategies for Handover and Dynamic Channel Allocation in Micro-cellular Mobile Radio Systems", *Proc Veh Tech Conf,* VTC-89, San Francisco, CA, 1989.

[3]   Everitt, D.E., MacFadyen, N.W., "Analysis of Multicellular Mobile Radio Telephone Systems with Loss", *BT Tech. Journal,* Vol 1 No 2, 1983.

[4]   Halpern, S.W., "Reuse Partitioning in Cellular Systems", *Proc Veh Tech Conf,* VTC-83, 1983.

[5]   McEliece, R.J., Sirarajan, K.N.,"Performance Limits for FDMA Cellular Systems Described by Hypergraphs", *Proc Third IEE Conf on Telecommunications,* Edinburgh, 1991.

[6]   McEliece, R.J., Sirarajan, K.N.,"Asymptotic Performance of Fixed and Dynamic Channel Assignment in Cellular Radio, *Proc IEEE Inf Theory Symp,* ISIT-91, Budapest, 1991.

[7]   Eriksson, H., Bownds, R., "Performance of Dynamic
        Channel Allocation in the DECT System, *Proc Veh Tech
        Conf,* VTC-91, St. Louis, MO, 1991.

[8]   Zander, J, Frodigh, M., "Capacity Allocation and Channel
        Assignment in Cellular Radio Systems using Reuse-
        partitioning" *Electronics Letters,* Vol 28,
        No 5, February 25, 1991.

[9]   Zander, J, Eriksson H., "Asymptotic Bounds on the
        Performance of a Class of Dynamic Channel Assignment
        Algortihms" *Internal Report,* TRITA-TTT-9120, ISSN-0280-
        4492, Royal Institute of Technology, Stockholm, Sweden

ON THE MAXIMUM CAPACITY OF POWER-CONTROLLED CELLULAR NETWORKS

Nicholas Bambos and Gregory J. Pottie

Department of Electrical Engineering,
University of California at Los Angeles,
Los Angeles, CA 90024

ABSTRACT

We are concerned with limiting the cochannel interference between transmitter-receiver pairs in a cellular network below acceptable levels, while accommodating as many transmissions as possible. The controllable network parameters are the transmission powers of the base stations, and possibly of mobiles too, in the cells. Efficient adaptive control of the powers is utilized to guarantee quality of service above a certain threshold at all times, given the continuously changing traffic of communication requests in the network. The quality criterion used in this study is the carrier to interference ratio (C/I) at the receiver.

1. Introduction

In a single cell radio system, any orthogonal decomposition may be used to divide up the available bandwidth efficiently, provided of course that adequate diversity is obtained. A central controller with a reservation-based access control protocol may then manage the resources most efficiently. However, in a multi-cell environment, the attenuation suffered by the signal over distance is insufficient to isolate the cells from each other. Consequently, the dominant channel impairment in conventional architectures is interference from nearby cells.

For Time Division Multiple Access (TDMA) systems, interference on the uplink, i.e, mobile to base station link, is due to transmissions by other mobiles sharing the same slot in nearby cells. In the simplest case, signal power is attenuated as the fourth power of distance. This partially isolates the cells, allowing some re-use of the slots or frequencies. Re-use is limited by the need to maintain some minimum carrier to interference ratio (C/I), the value of which depends on the modulation and diversity techniques employed. The worst case condition occurs when the mobile is near the cell boundary, and a mobile in a neighboring cell is also near the cell boundary and transmitting in the same slot. In that situation, C/I is approximately unity and reliable communication is impossible.

The traditional solution is to assign a fixed band of frequencies to each cell, with the pattern arranged so that the worst case interference conditions result in acceptable C/I. Typical re-use patterns result in efficiencies of 1/4 to 1/7. This approach has the virtue of simplicity, but is suboptimum in terms of efficiency.

In Code Division Multiple Access (CDMA) systems, by contrast, the frequency re-use is determined by average interference conditions rather than the worst case [1]. Roughly speaking, the de-spreading of the desired signal spreads the interferers so that no one interferer dominates. Unfortunately, it is very difficult to maintain synchronous operation in the uplink resulting in the generation of interference within the cell. This interference actually dominates for a fourth-power attenuation law.

The incentive for more creative management of power and slot allocation in TDMA is that we do away with the interference generated within the cell, and can potentially reduce worst-case interference to the average level of the out-of-cell interference. That is, situations where the C/I would be unacceptable are avoided through access and power control. In section 2 we give a simple example which illustrates the main features of the problem, including both the potential benefits and drawbacks of dynamic control. In sections 3 and 4 we more formally describe the optimum resource control problem, while in section 5 we indicate topics of further research.

2. Motivation for Dynamic Power and Access Control

In the following example, it is assumed that attenuation obeys a fourth power law. It is well known that performance of CDMA systems is improved if power is controlled such that the received signal levels at the base station are all approximately equal. Normalize the cell radius to 1 and assume that the cells have been divided into three equal sectors, as suggested in [1]. Let $\phi(r)=\alpha r^4$ where α is a constant determined by the average power constraint:

$$\int_0^{2\pi/3} \int_0^1 \phi(r)\, r\, dr\, d\theta = 1$$

Let η denote the fraction of the codes that are active, i.e., the efficiency of use of the available spectrum. Then the in-cell interference is given by

$$I_{in} = \frac{2\pi}{3} \int_0^1 \eta \frac{\phi(r)}{r^4} r\, dr = \frac{\pi\alpha\eta}{3}$$

Because of the fourth power law, and the dependence of power with distance from the base station, most of the out-of cell interference comes from the boundary. A simple and reasonably accurate overbound on the interference from one of the two neighboring sectors is as follows:

$$I_{ex} = \int_0^1 \frac{\alpha r^5}{(2-r)^4}\, dr = \alpha/6$$

where r is the radius from the neighboring sector. Thus, since two sectors are in view, the total interference is approximately $1.3\alpha\eta$, and the received signal level is α. Thus, C/I is approximately $3/4\eta$. That is, it is inversely proportional to the efficiency.

For the downlink, i.e., base station to mobile, it is desirable that C/I be independent of r. The interfering base stations are at distance 2, and transmit unit power. The distance to the mobile is (roughly) 2-r, where r is its distance from its own base station. In this case, synchronous operation is feasible, and so we will assume there is no in-cell interference. This implies that

$$\frac{\phi(r)}{r^4} / \frac{2\eta}{(2-r)^4} \quad C/I$$

that is,

$$\phi(r) = \frac{2\eta r^4}{(2-r)^4} \quad C/I$$

Applying the average power constraint, we obtain $C/I = 1.4/\eta$. Note that the numbers work out to be much the same in a non-sectorized system, and thus capacity increases linearly with the number of sectors.

If the efficiency is set to 1/3, C/I is 3.5 dB for the uplink and 6.5 dB for the downlink (which are pessimistic due to the approximations). For a TDMA system we may assign a frequency re-use of 1/3 to the sectors, such that we need consider only next-nearest neighbors. Let us suppose for the downlink that $\phi(r)$ is the same as in CDMA, and we assign the same slot to mobiles at the same radius. Thus, for a mobile at radius r, the interfering signal travels a distance of 2+r. Then the interference power due to two interfering cells is given by

$$I = \frac{2\phi(r)}{(2+r)^4}$$

The signal power is $\phi(r)/r^4$, and thus C/I is minimized for $r=0$, to the value 16 dB. If the mobiles are not constrained to have the same slot at the same radius, the minimum C/I is 9 dB. The same results may be obtained for the uplink, except that the optimum solution leaves $\phi(r)$ free. Even in this simple example, it pays to exercise control over slot assignments. The extra margin could for example be used to increase the number of levels of modulation, reduce the redundancy of the error correcting code, or improve frequency re-use among sectors, to increase the capacity.

In a real system, the positions of the mobiles are not known, and the propagation characteristics are much more complicated. Superimposed on the attenuation with distance are random fluctuations due to shadowing and ducting. These have relatively little impact on the capacity of CDMA systems with power control, but they will very adversely affect the worst case interference conditions for a conventional TDMA system. For example, log-normal fading with a variance of 8 dB would cause the 90% worst-case C/I of TDMA systems to drop well below that of CDMA systems, unless the slots and power are dynamically assigned.

For optimum dynamic power and access control, the base stations must have up to date information on the set of gains between base stations and mobiles. Since the base station controls the transmit powers of its own mobiles, assuming rough reciprocity in the propagation characteristics it can estimate the G's. (This would be true for example in time division duplex systems). However, the gain between a base station and a mobile in a neighboring cell is more difficult to determine, since all the base station can measure is the total interference in any given slot. The neighboring base station knows the transmit power level, but the mobile must measure the gain. Mobiles must in any case determine which base station will provide the strongest signal to allow efficient hand-offs between cells. Base stations might transmit pilot tones for this purpose. The size of the set of tones would be similar to the frequency re-use patterns of conventional systems, with a larger set giving better accuracy in computing the gain to a particular base station. Mobiles could monitor the strength of the pilots in each frequency, and report this information periodically to their own base station. Thus, optimal control requires intelligent mobiles, exchange of control information between mobiles and base stations over the radio link, and exchange of information between base stations using the backbone network.

If it is not possible for base stations to exchange information on gains, assignments, and power levels, it may still be possible to do better than a fixed assignment system. Base stations can make allocation decisions for the uplink based on total measured interference for each slot. Smart mobiles could also monitor interference and pass a short list of desired slots for the downlink to the base station along with their access requests. Requests for access would be delayed if it appeared that no slot would provide acceptable C/I consistent with the average power constraints.

A number of schemes that use such local information to assign slots, power or both have previously been proposed, e.g. [2], [3]. Zander [5] has considered power control in cellular nets, studying the maximum capacity, using the theory of positive matrices. We take here a different approach based on constructing an optimal solution (when it exists) for any required quality of service and approximating it in a computationally efficient manner. The resulting control scheme is both dynamic and adaptive. In the following we describe the allocation problem when base stations may interact.

3. Call Admission Control in a Single Channel.

We start with a fixed geometric structure of the cellular network, essentially specified by the fixed positions of the base stations; two distinct base stations may have the same position, transmitting, however, in disjoint solid angles through highly directional antennae. We consider only cochannel interference, and focus on the downlink problem, that is, base-to-mobile communication. The uplink problem can be essentially decoupled from the downlink one (mobile-to-base communication). It can be treated analogously, at least at the algorithmic level of admission control, which is examined in this section.

Suppose there are N base stations, indexed by $i \in C=\{1, 2, 3, ..., N\}$. Let P_i be the (average) power transmitted by base station $i \in C$, in the channel (slot) under consideration. $P_i = 0$ means that the i-th base station is idle in that channel. Let $C_a \subset C$ be the set of all *active* (non-idle) base stations at a given time, that is, $P_i > 0$ for every $i \in C_a$ and $P_i = 0$ for every $i \in C-C_a$. Each base station can transmit to at most one mobile in any given slot; similarly each mobile can transmit to at most one base station. The power received by the mobile, to which the i-th base station is transmitting, is given by $C_i = G_{ii}P_i$ (carrier power), where G_{ii} is the power gain between the base station and the mobile, and P_i is as defined previously. However, the same mobile is receiving power $I_i = \sum_{\{j \in C_a-\{i\}\}} G_{ij}P_j$ (cochannel interference power) from the rest of the base stations (excluding the i-th one), which is naturally treated as interference; G_{ij} is the power gain between the j-th base station and the mobile with which the i-th base station is communicating directly. The power gains between the base stations and the mobiles are estimated dynamically as described in the previous section.

In view of the above, the carrier to interference ratio at the mobile receiving from the i-th base station is given by:

$$R_i = \frac{G_{ii}P_i}{\sum\limits_{j \in C_a-\{i\}} G_{ij}P_j},$$

for every $i \in C_a$. The quality of each transmission is guaranteed in the sense that $R_i > \gamma$, $i \in C_a$, where γ is a lower bound reflecting the lowest acceptable global service quality of the network, and is given as a required specification. This set of conditions is equivalent to:

$$D_i = G_{ii}P_i - \gamma\{\sum_{j \in C_a-\{i\}} G_{ij}P_j\} > 0,$$

for every $i \in C_a$. Essentially, proper operation of the network corresponds to the existence of a power vector $\vec{P} = (P_1, P_2, P_3, \cdots P_N)$, such that $D_i \geq 0$ for every $i \in C_a$ and $P_i = 0$ for every $i \in C-C_a$. That is the linear inequalities above are satisfied by the powers of all the active base stations, given the constants G_{ij}. Such a power vector is called C_a-**feasible**. The base station powers just have to form a feasible

$\vec{P}$, in order for the network to provide acceptable quality to all users of the channel. It is very important to note that any positive scalar multiple of a feasible power vector is also feasible. This implies that it is enough to always choose $\vec{P}$'s that have all their components below the maximal power that a base station can transmit. It also implies the set of all C_a-feasible power vectors, that is, the **feasibility region** is always an infinite (unbounded) *polyhedral cone* with tip at $\vec{0}$ point.

The problem is that as the number of transmissions increases the consistency relations may be violated, in the sense that there may not exist a feasible power vector to accommodate all the active base stations. In view of this, when a new request for transmission appears, the network has to exercise admission control; it will reject the request if there is no feasible power vector that accommodates it, otherwise, it will accept it and readjust the powers at the active base stations according to the new feasible vector. The operation of checking whether the a new transmission request can be accommodated has to be done in real time, therefore, it must have low computational complexity.

The simplest possible admission control scheme (clearly suboptimal) would be the following. Given the set of active stations C_a, suppose that a previously idle station $s \in C{-}C_a$ wants to become active by accepting a call. The call is accepted only if there is a positive $P_s > 0$ (which will be the power of the s base station), such that $\min_{i \in C_a} \{D_i - \gamma\, G_{is} P_s\} = G_{ss} P_s - \gamma\{\sum_{\{j \in C_a\}} G_{sj} P_j\}$; if there is no such positive P_{n+1}, the call is rejected. Using the above admission criterion the inequalities are always fulfilled for the active transmissions. However, it is easy to see that the policy is not optimal, since it takes no action to adapt to the changing environment.

In order to construct the optimal (highest throughput) policy, we start by studying the necessary and sufficient conditions for the existence of a feasible power vector for a given set of active stations $C_a \subset C$. The following theorem is the key point of the optimal admission scheme.

Theorem

For an $m{\times}m$ invertible matrix $\mathbf{A}$ with positive diagonal elements and negative off-diagonal ones , if there exists a componentwise positive vector $\vec{x}_o$, such that $\mathbf{A}\vec{x}_o$ is also positive componentwise, then the equation $\mathbf{A}\vec{x} = \mathbf{1}$ has exactly one non-zero solution, and that is also positive componentwise. The vector $\mathbf{1}$ has all its components equal to 1.

Proof:

Let $\mathbf{R}_+^m$ be the set of all componentwise positive vectors of the m-dimensional real space. Let also the rows of the matrix $\mathbf{A}$ be given by the m-dimensional vectors $\vec{a}_1, \vec{a}_2, \cdots \vec{a}_\mu, \cdots \vec{a}_m \in \mathbf{R}^m$. The inequality $\vec{a}_\mu \vec{x} > 0$ is fulfilled on just one of the two m-dimensional half-spaces separated by the m-dimensional hyperplane $\vec{a}_\mu \vec{x} = 0$ (passing through $\vec{0}$), for every $\mu \in \{1, 2, \cdots m\}$. No two such hyperplanes are parallel (identical since they have a common point $\vec{0}$), because the matrix $\mathbf{A}$ is invertible. Multiplication of vectors is understood under the usual inner product.

Define now $S_\mu = \{\vec{x} \in \mathbf{R}^m : \vec{a}_\mu \vec{x} > 0\}$, $\mu \in \{1, 2, \cdots m\}$, and $S_* = \bigcap_{\mu=1}^m S_\mu$. The existence of a positive vector $\vec{x}_o \in \mathbf{R}_+^m$ fulfilling all the inequalities $\vec{a}_\mu \vec{x} > 0$, for every $\mu \in \{1, 2, \cdots m\}$, guarantees that $S_* \neq \varnothing$ (is non-empty). Therefore, S_* must be a polyhedral cone, extending to infinity (unbounded), with tip at $\vec{0}$, being a connected and convex set.

We now prove that $S_* \subset \mathbf{R}_+^m$. Since the element $\vec{x}_o \in \mathbf{R}_+^m$ belongs to S_* we conclude that $S_* \cap \mathbf{R}_+^m \neq \varnothing$. Also S_* is a connected and convex set. Arguing by contradiction, suppose now that a non-zero point $\vec{b}$ of the boundary of the convex set $\mathbf{R}_+^m$ belongs to the set S_*, that is the boundary intersects S_*. This point $\vec{b}$ must have at least one (but not all) component equal to zero, while its non-zero components will be positive. Putting this vector b into the inequalities $\vec{a}_\mu \vec{b} > 0$, we can see that at least one will be violated, because the off-diagonal elements of the matrix $\mathbf{A}$ are negative, leading to a contradiction. Therefore, $\vec{b}$ cannot belong to S_*, and S_* has to be a subset of $\mathbf{R}_+^m$.

Consider now the sets $V_\mu = \{\vec{x} \in \mathbf{R}^m : \vec{a}_\mu \vec{x} = 1\}$, for every $\mu \in \{1, 2, 3, \cdots m\}$. Obviously $V_\mu \subset S_\mu$ for all μ's, and also $V_* = \bigcap_{\mu=1}^m V_\mu = \{\vec{x}_*\}$, where $\vec{x}_* \neq 0$ is the unique solution of the system $\mathbf{A}\vec{x} = 1$ (due to the invertibility of $\mathbf{A}$). Therefore, $\vec{x}_* \in V_* \subset S_* \subset \mathbf{R}_+^m$. This completes the proof of the theorem.

Remark: The theorem still holds if not all the off-diagonal elements of $\mathbf{A}$ are negative, but instead we require that on every row of $\mathbf{A}$ some of the off-diagonal elements (but not all) are zero and the rest are negative.

In light of the previous theorem we can examine the problem of existence of a feasible power vector for a given set of active stations. Indeed, suppose that we are given a set of active stations $C_a \subset C$ and a

C/I threshold γ, and we want to now whether there is a set of powers $P_i > 0$, $i \in C_a$, such that the standard **feasibility relations**

$$G_{ii}P_i - \gamma\{\textstyle\sum_{j \in C_a-\{i\}}G_{ij}P_j\} > 0$$

hold for every $i \in C_a$. We can write the above relations in matrix form

$$G\vec{\Pi} > \vec{0},$$

where G is the matrix of the multiplicative elements of the feasibility relations (depending on the power gains and γ), and $\vec{\Pi}$ is the vector of powers P_i ($|C_a|$ of them), renumbered appropriately. The key point is that the matrix G has positive diagonal elements and negative off-diagonal ones (they match actually the relaxed requirements of the remark above); also G is invertible, since even if its not, still any infinitesimal perturbation (expressing geometrical and physical asymmetries etc.) of the natural parameters G's would actually make it invertible (thus, we can safely assume its invertible). Therefore, in view of the previous Theorem, if there exists a set of positive powers fulfilling all the compatibility relations, the solution of the equation $G\vec{\Pi} = 1$ has a positive solution.

We thus try to solve the equation $G\vec{\Pi} = 1$, which we call **feasibility equation**. It can be solved by Gaussian elimination, which is a very well understood process, and naturally amenable to parallel computation. Indeed, we can find the solution in $O(N^2)$ elementary steps on N parallel processors, or in $O(N)$ steps on N^2 parallel processors. The computational complexity is rather low, allowing the solution of large sets of base stations in very small time. This leads to efficient admission control schemes, implementable in real time. The complexity of the Gaussian elimination problem can be further reduced in our situation, due to the highly localized structure of interactions of the stations, leading to very sparse matrices G. Solving the equation $G\vec{\Pi} = 1$ we identify a solution $\vec{\Pi}_*$. If this solution is all-positive componentwise, then this generates a C_a-feasible power vector, and by setting the powers of the base stations according to the components of the vector Π_* we can guarantee that the C/I ratios at all active stations are above γ, as explained above. However, if the solution of the above equation is not all-positive (even one component is not positive), then, according to the previous Theorem, this implies that there is no feasible power vector!

The **optimal admission control policy** then works as follows. Given the set of active stations C_a (fulfilling the feasibility relations), suppose that a previously idle station $s \in C-C_a$ wants to become active by accepting a call. The G_{ss}, G_{si}'s and G_{is}'s ($i \in C_a$) are estimated, and the feasibility equation is solved for the set $C_a \cup \{s\}$. If there is an all-positive solution, the powers are adjusted according to that solution and the new call is admitted. If there is not an all-positive solution, then the call is rejected and the powers remain as they were.

In this section we have examined very briefly the theoretical fundamentals of the optimal admission scheme. We have not elaborated on the implementation details. Current research is focusing on reducing the computational complexity of the admission scheme, exploiting the local nature of the problem, and on implementing it in a distributed and asynchronous manner.

4. Call Admission Control over Multiple Channels.

When the feasibility equation is solved and an all-positive componentwise solution results, then each of the feasibility relations will have the form

$$G_{ii}P_i - \gamma\{\sum_{j \in C_a-\{i\}}G_{ij}P_j\} = q$$

where q, the **quality index**, is a positive constant. We assume that there is a common lower bound for the powers transmitted by the stations. Large q indicates the likelihood that new calls can be admitted in that slot, while q approaches zero as the slot fills up.

The quality index can be used to help achieve load balancing in a multi-channel (slot) system. One possible allocation scheme is as follows.

Algorithm 1

- Produce a list of slots for which the base station has currently assigned no calls. For the slot with the highest index q, solve the feasibility equation to see if the call can be admitted. If successful, compute the new quality index.

- If unsuccessful, proceed through the list of remaining slots until either the call is admitted or all slots have been tried. If unsuccessful, the call is blocked.

We have performed a simulation of this algorithm, for channel conditions described in [1]. A 43 cell network consisting of 3 concentric rings of hexagonal cells about the central cell was used, with statistics reported only for the center cell to avoid edge effects. Power attenuation with distance was assumed to obey a fourth power law, and mobiles were randomly generated according to a uniform geographic distribution. Log-normal shadowing with a standard deviation of 8 db was assumed, with independent shadowing for each track. Each cell was divided into three 120 degree sectors. The simulation procedure was to start with no calls, and add calls until the blocking probability exceeded 2% for the innermost cell. This was repeated until at least 25 blocked calls were reported. Perfect knowledge of the set of gains G was assumed, the efficiency of slot usage in each sector, n, is plotted in Figure 1 as a function of the threshold γ, and 64 slots were available to each sector.

The high efficiency can be explained as the combination of two effects, namely, statistical multiplexing and optimum power control. As previously noted, in a conventional TDMA system we are limited by worst case interference conditions. However, consider a situation where received power levels for all mobiles in a cell are controlled to be equal (not optimal) and mobiles are assigned to slots for minimal interference. With a very large number of slots, matters can be arranged such that the signal to interference ratio is approximately equal in all slots. For the situation described above, the interference power I is then related to the desired signal power C and the fraction of slots occupied in every cell n by $C/I = 1/0.65n$. This is also illustrated in Figure 1. Efficiency is lower than that found in the simulation because the power control scheme is not optimal; with the procedure of algorithm 1, the interference levels are lower than those obtained by constraining received power levels to be equal. Thus, both slots and power must be controlled to achieve optimum efficiency.

The results may also be compared to those obtained for DS/SS CDMA in [1]. For a 1% probability that the bit error rate exceeds 0.001, 36 calls can be accommodated in a 1.25 MHz band, for 8kb/s voice, and signal to interference power of 7dB. This included a voice activity factor of 3/8, which could of course also be exploited in a packet-based TDMA system. Neglecting, this, in our terms an efficiency of 10% is achieved at $\gamma = 7$dB. For comparison, our earlier example gave an efficiency of 33% at $\gamma = 3.5$dB. Both these points lie about 5dB to the left of the curve predicted by the statistical multiplexing model outlined above, because of course the in-cell interference is larger than the out-of-cell interference. Efficiency as a function of γ is even further inferior to that of algorithm 1.

Caution must be exercised in interpreting these results. First, optimum slot and power allocation demands more complexity in the network. Second, implicit in this comparison is the assumption that signal to interference ratio is an accurate predictor of error performance. In CDMA, it is reasonable to assume the residual interference following the despreading operation has Gaussian statistics. For dynamic slot/power control, as for conventional systems the interference will be dominated by a small number of users, and consequently the Gaussian assumption is not valid. Thus, while the average signal to interference ratio is improved, its character is not, and a higher γ is required than for CDMA to achieve the same error performance for 10% and 1% worst case situations. For example, in conventional systems $\gamma = 18$dB is assumed necessary for adequate performance, which for algorithm 1 results in an efficiency of 30%. This still represents a three-fold improvement in capacity with respect to CDMA with $\gamma = 7$dB, even though the conventional system does not include either diversity or coding.

It is possible to improve upon algorithm 1, using a variant of a method proposed in [3]. By re-assigning a call that has previously been accepted more calls might be fit in. Therefore the following procedure may be attempted:

Algorithm 2
- Invoke Algorithm 1.
- On failure to slot a call, according to some arbitrary procedure pick a slot which already has a call assigned. Remove that call, and solve the feasibility equation for the new call.
- If successful, attempt to admit the previously accepted call in another slot.
- If unsuccessful in re-slotting, restore the previous situation. The new call is blocked after all previously occupied slots have been tried.

This algorithm involves much more computation than algorithm 1 in conditions of heavy use, but leads to significant performance improvement as illustrated in Figure 1. Further gains, might be obtained by allowing more calls to be moved to accommodate new ones, as discussed in [3]. However, this is a case of diminishing returns for dramatically increased effort.

5. Final Remarks

We have studied the optimal power and slot allocation problem for cellular networks. Further research will include new algorithms exploiting the idea introduced in sections 3 and simulation of the performance of specific schemes under restrictions on the quantity and quality of channel state information transferred among mobiles and base stations.

6. References

[1]   K. S. Gilhousen, et al., "On the Capacity of a Cellular CDMA System," IEEE Trans. Vehic. Tech., vol. 40, no. 2, pp. 303-312, May 1991.

[2]   H. Panzer and R. Beck, "Adaptive Resource Allocation in Metropolitan Area Cellular Mobile Radio Networks," IEEE Vehic. Tech. Conf., May 6-9, 1990, Orlando FL, pp. 638-645.

[3]   K. N. Sivarajan, R. S. McEliece and J. W. Jackson, "Dynamic Channel Assignment in Cellular Radio," IEEE Vehic. Tech. Conference, May 6-9, 1990 Orlando FL, pp. 631-637.

[4]   D. P. Bertsekas and J. N. Tsitsiklis, "Parallel and Distributed Computation - Numerical Methods," Prentice Hall.

[5]   J. Zander, "Performance of Optimum Transmitter Power Control in Cellular Radio Systems," IEEE Trans. Vehic. Tech., vol. 41, no. 1, pp. 57-62, Fed. 1992.

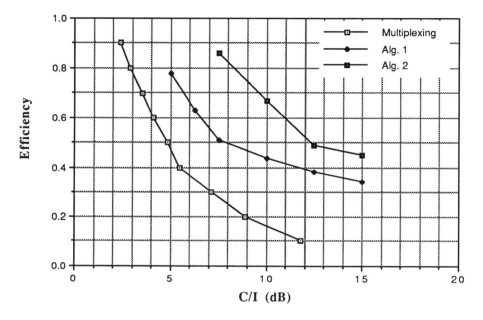

Figure 1: Slot Efficiency vs. C/I threshold

Multimedia Personal Communication Networks (PCN): System Design Issues

D. Raychaudhuri and N. Wilson
David Sarnoff Research Center
Princeton, NJ 08543-5300

Abstract

This paper presents a discussion of system design issues related to the development of next-generation multimedia personal communication networks (PCN). It is proposed that the baseline definition of emerging PCN systems be broadened to provide multimedia transport capabilities qualitatively similar to those offered by BISDN/ATM. A number of system-level design and technology selection issues are discussed. Specific topics considered include: multimedia PCN traffic handling requirements, service cost targets, system design criteria, choice of packet switching vs. circuit switching, strategies for ATM compatibility, media access control (MAC) selection, etc. The presentation is concluded with an outline of a dynamic TDMA based multimedia PCN system architecture consistent with the preceding discussion. Simulation based performance evaluation results are also given for such a dynamic TDMA PCN channel operating in example mixed voice/data traffic scenarios.

1. INTRODUCTION: Personal communication networks (PCN) based on emerging digital wireless technologies are expected to play a significant role in next-generation telecommunication systems. While some general concepts for PCN have been proposed by many authors [GO90, GO91, CO87, CO90, ST89, ST90], broad consensus on a common conceptual framework for such systems is still several years away. Clearly, the evolution of a widely accepted set of PCN system designs and associated technologies will be influenced by results from research/development, prototyping and field trial activities being conducted by various organizations worldwide. This paper, which is intended as another input to the ongoing discussion on PCN system architecture, proposes that next-generation PCN systems be designed to provide integrated multimedia transport capabilities qualitatively similar to those offered by broadband ISDN/ATM.

Consider first the definition of "personal communication network", often used synonymously with the phrase personal communication system (PCS). It is generally agreed that PCN will provide reliable, ubiquitous communications services to individuals via small, portable, low-power terminals operating in a "microcell" environment. In view of the growing importance of data and multimedia (i.e., integrated voice, video/image and data) in the wired telecommunication network, we believe that future PCN systems should be be capable of supporting such advanced multimedia services. If "third generation" PCN systems are designed to provide flexible transport services for voice, video and data along the lines of BISDN/ATM, then separate (an possibly incompatible) technologies for digital cordless telephone and wireless local area network (LAN) would no longer be required in the long run. On the other hand, a complete merging of microcellular PCN with (high power, large cell) cellular mobile seems unlikely, although there could be some areas of overlap. These expectations for evolution of wireless communication technologies are illustrated schematically in Fig. 1.

290

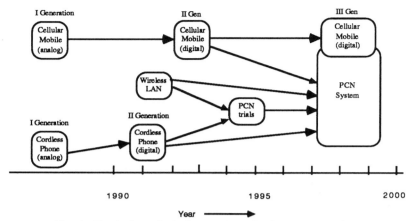

Fig. 1. Evolution of wireless communication technologies.

In the rest of this paper, we present a discussion of multimedia-capable PCN system design issues, in order to provide some understanding of functional requirements and applicable technology selection trade-offs. The discussion is concluded with an illustrative example of a multimedia PCN architecture that is consistent with the system design criteria discussed.

2. PCN SYSTEM ISSUES

PCN Traffic & Its Requirements: PCN networks deployed 5-10 years from now will have to support not only digital voice communications, but also a reasonable range of data/multimedia traffic. The data traffic will be composed of a mixture of conventional computer communications and electronic mail, along with data generated by a variety of emerging digital business and consumer applications. Examples of these new applications are video teleconferencing, portable PC-based multimedia voice/video/data applications, digital TV/audio delivery, information distribution/ retrieval services, etc. A summary of potential data sources for PCN is given in Table 1.

Table 1. Examples of PCN traffic sources and their performance requirements

Application	Data type	Avg. Data rate (Kbps)	Peak Data rate (Kbps)	Max.Delay (sec)	Max.Pkt Loss rate
e-mail, paging	VBR	10^{-2}-10^{-1}	10^0-10^1	$<10^1$-10^2	$<10^{-9}$
Computer data	VBR	10^{-1}-10^0	10^1-10^2	$<10^0$-10^1	$<10^{-9}$
Telephony	CBR	10^1-10^2	10^1-10^2	$<10^{-1}$-10^0	$<10^{-4}$
Digital audio	CBR	10^2-10^3	10^2-10^3	$<10^{-2}$-10^{-1}	$<10^{-5}$
Teleconference	CBR/VBR	10^2-10^3	10^3-10^4	$<10^{-3}$-10^{-2}	$<10^{-5}$

While it is difficult to predict precisely which applications will generate significant traffic demands on PCN, it is clear that a good system must be capable of handling a variety of two-way and one-way digital transmission needs with a range of bit-rates and performance requirements. The consumer of PCN 10 years from now may also be receiving flexible ATM-based broadband ISDN service over the wireline network, and can thus be expected to have a variety of digital devices, of both connection-oriented and connectionless types. PCN networks of the future must thus be designed to provide flexible remote access services that complement the B-ISDN network, rather than just servicing voice calls. Of course, there will be certain broadband applications (e.g., HDTV, high-speed workstation file transfer, etc.) that cannot be viably supported on PCN due to inherent bit-rate and/or service quality limitations of the shared wireless medium, but it is hoped that these restrictions will not apply to the majority of popular consumer-level digital services. Fig. 2 shows our assessment of the target traffic regime (characterized by two attribute ranges: average bit-rate and the % of data traffic) appropriate for a prototype PCN system.

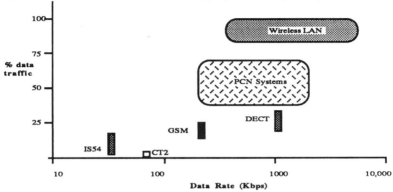

Fig. 2: Target traffic regime for PCN

A related issue is that of developing estimates for total volume of traffic (in an appropriate unit such as bps/Km^2) that needs to be supported by future PCN. To obtain a rough estimate, consider a nominal urban /suburban population density of 1000 people/Km^2. Assuming 25% penetration of the PCN service, there would be 250 subscribers for voice and other data. Assuming voice to be 16 Kbps ADPCM at 0.2 erlangs/subscriber, the loading due to voice would be 800 Kbps/Km^2. While data traffic volume is more difficult to estimate, if we assume 50% data, then the total spatial throughput required is approximately 1600 Kbps/Km^2. This could, for example, be achieved with 4 cells / Km^2 (i.e., cells of approx. 0.5 Km) each with full duplex transmission rates of ~1 Mbps and ~50% transmission efficiency.

PCN Cost Targets: Since PCN technology is at a relatively early stage, achievable cost levels are difficult to estimate. However, it is helpful to consider general network and terminal equipment cost ranges which would serve as a guide to technology selection, etc. It is generally accepted that PCN terminal cost/complexity should be considerably below that of present-day cellular mobile, typically by an order of magnitude. Transmission/network cost has not been discussed as much - it is our opinion that in order to become a truly universal service with wide acceptance, PCN network usage costs (measured in an appropriate unit such as $/Kb) must be well below that of cellular mobile. In addition, it would be desirable to provide a even lower cost variable bit-rate (VBR) service for packet data users sharing the network. An approximate assessment of the cost targets for PCN as compared with cellular mobile, wireless LAN and cordless telephone is shown in Fig. 3.

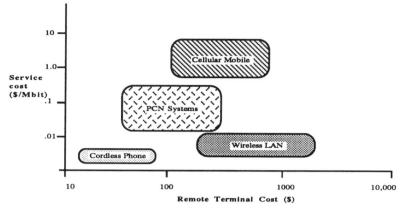

Fig. 3. Approximate cost targets for PCN

It is observed that these lower cost goals may be achieved through an appropriate system design which includes: low power/complexity subscriber terminals; efficient spectrum usage via proper selection of cell size, modulation, medium access control, etc.; dynamic sharing of bandwidth via techniques such as demand assignment, packet switching, asynchronous multiplexing, etc.; effective sharing of network hardware resources such as base station receivers, switches, backbone network, etc.

PCN System Design Criteria: The design of a PCN system involves making judicious trade-offs between a number of performance, cost and equipment size/complexity factors. The primary design criteria for PCN systems include:

(a) <u>System Capacity (per unit bandwidth)</u>: PCN capacity is best expressed in a suitably normalized unit such as S = Throughput (Kbps) per unit area (Km^2) per Mhz of spectrum. Thus S is a measure of the bandwidth efficiency of a PCN system, and can be thought of as the single most important system performance measure. Systems with larger values of S are the primary aim of many research activities on this topic, given the scarcity of bandwidth. There are two basic approaches to increasing capacity: the first is to improve the bandwidth efficiency of key subsystems such as the modulation and medium access control, while the second is to decrease cell size for increased spatial re-use.

(b) <u>Subscriber Performance</u>: In previous wireless systems intended solely for voice (e.g., cellular mobile), performance measures are relatively simple and can be expressed in terms of familiar quantities such as voice signal-to-noise ratio (for analog systems) / bit-error rate (for digital systems) and circuit blocking probability. For newer integrated systems such as PCN, it will be necessary to set performance specifications for other services such as packet data (datagram type), connection oriented VBR or CBR data service for multimedia, video, etc. as well as for digital voice. Some guidance can be obtained from the service criteria used in designing B-ISDN. However, it is important not to overspecify performance on PCN since the wireless medium has inherent limits which are well below those possible with fiber media. For example, for data services, it may be realistic to specify wireless network response times of the order of 10's of ms, thus providing sufficient margin for propagation time, media access delay, etc. Similarly, for CBR services, nominal bit-error rates should be specified in the range of 10^{-5} to 10^{-7} rather than the 10^{-9} regime associated with fiber systems, given the difficult transmission properties of wireless media. Applications such as multimedia would still work over such a network, although some additional effort may have to be given to the subject of robust design. Another performance measure of interest for many new services is the peak transmission bit-rate, since advanced multimedia, video and computer graphics applications may require transfer rates of the order of at least 100's of Kbps (or preferably Mbps). This peak bit-rate capability may be a major distinguishing factor between different PCN systems, even though standard voice can be supported even at 10's of Kbps.

(c) Remote terminal cost/complexity: Since PCN is planned as a mass communication service, remote terminal cost and complexity are of vital importance, and may sometimes be more significant than capacity. In Fig. 3 above, we have outlined the range of subscriber terminal cost that may be acceptable for PCN. While the cost /complexity may vary depending on service requirements, a low-end unit supporting voice only and/or e-mail will have to be in the same price range as cordless phones, while more advanced units attached to multimedia PC's etc. may perhaps cost up to 10% of the device supported.

(d) Remote terminal power level & battery consumption: The power level transmitted by the remote terminal has a direct impact on battery consumption, a critical element for portable equipment design. A general goal that has been discussed by PCN proponents is to maintain battery consumption at ~100 mW when active, dropping to ~1-10 mW when inactive. This corresponds to relatively low remote transmit powers (~10's of mW) rather than the higher (~1 W) power levels associated with cellular mobile. Since battery technology cannot be expected to follow the general speed/processing curves characteristic of electronic components, it is important to design the system to take advantage of signal processing/computing power to reduce battery consumption. For example, this would mean giving careful consideration to advanced error correction techniques that lower power level requirements or medium access techniques that result in a low transmit duty cycle.

(e) Regulatory implications: Since PCN systems will require bandwidth allocations from national and international bodies, it is important that regulatory issues be considered in the design. The U.S. FCC has already issued some general guidelines/preferences in their recent notice of inquiry on PCS. It is clear that support from regulatory agencies will require systems to be based on low power levels (<<100 mW), and be capable of providing high spectral utilization. In order to justify allocation of new bands to PCN, proponents must demonstrate high bandwidth utilization along with an ability to serve mass communication needs (rather than just special purpose applications for a few people). In addition, the ability to operate in an "overlay" mode in existing microwave bands would be an important benefit, and spread spectrum methods have lately been proposed as a technical solution for PCN overlay.

Packet vs. Circuit Switching for PCN: In view of the unique, multi-service environment of multimedia-capable PCN, we believe that PCN systems should be based on resource-shared packet switching technology rather than circuit switching currently employed in cellular mobile systems.

There are several technological factors that tend to favor the use of packet transport (or fast-packet, cell-relay, etc.) techniques for PCN. These include:

(a) PCN will be required to serve a wide range of services including voice, data, e-mail, digital video, multimedia, etc. As for B-ISDN, separate circuit-switched channels cannot provide the necessary flexibility, especially because the bit-rates required by new services cannot even be reliably predicted.

294

(b) PCN systems will support large populations of infrequent (bursty) users, so that economic viability will depend on the ability to effectively resource share the bandwidth and infrastructural equipment such as base station transceivers, switches, etc.

(c) The relatively unreliable wireless communication channel used for PCN is better suited to packet switching than circuit switching. Packet switching does not require continuous operation at very low bit-error rates and can compensate for occasional errors with ARQ-based retransmission strategies employed at the data-link level. In addition, packet transport makes it possible to apply diversity techniques to improve reliability of reception.

(d) Large PCN systems with small cell size inevitably require a high-capacity switching infrastructure for routing of traffic between cells, etc. As the number of cells goes up, the switching problem can become unmanageable unless packet (or fast-packet) based solutions such as ATM switching, high speed metropolitan area network (MAN) are employed.

ATM compatible PCN interface format: A key issue in the design of PCN systems is the selection of a suitable data format for interfacing to other networks, both wireline and wireless. As mentioned above, PCN will have to interface to conventional circuit switched digital networks, future ATM/B-ISDN networks, second or third generation digital cellular systems, etc. To facilitate interoperability with ATM and to simplify the difficult problem of switching data to and from base stations of small cells, we recommend adoption of an ATM compatible fixed length cell-relay format for PCN.

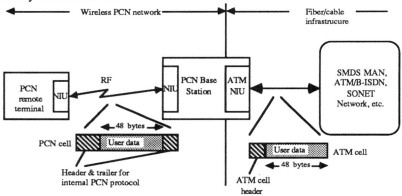

Fig. 4. ATM-compatible PCN approach

In this concept, the 48 byte ATM cell payload would become the basic unit of data within the PCN network, resulting in a transparent (or near-transparent) interface to an ATM (or SONET) based broadband backbone network fabric. Within the PCN, wireless channel specific protocol layers (e.g., medium access control layer) will be added to the ATM data unit as required, and replaced by ATM headers before entering the backbone network. This approach (illustrated in Fig. 4) has the benefit of being consistent with the expected availability of a broadband ATM infrastructure at about the same time as PCN systems start being deployed. The use of ATM switching for inter-cell traffic also avoids the crucial problem of developing a new backbone network with sufficient throughput to support intercommunication among large numbers of small cells.

It is noted that for small PCN cells with relatively low traffic volumes rather than direct connection to an ATM switch, it may be appropriate to use a resource-shared ATM-compatible IEEE 802.6 "SMDS" MAN to interconnect several base stations. These IEEE 802.6 MAN's can then be interconnected with an appropriate ATM switching arrangement.

Media Access Control (MAC) Selection: A key technical issue related to multimedia PCN design is the selection of a suitable channel sharing / media access control (MAC) technique at the data link layer (for inbound remote-to-base transmissions). The MAC technique used in PCN will have a significant impact on user performance, system capacity, and to some extent remote terminal complexity, and is therefore an important design parameter. A variety of MAC alternatives exist for PCN; major options include: CBR mode TDMA similar to that used in digital cellular (such as GSM), dynamic TDMA supporting both CBR and VBR modes of operation, carrier sense multiple access (CSMA) based techniques similar to those used in some wireless LAN's, circuit mode CDMA (similar to that recently proposed for digital cellular), packet CDMA [RA81a,86], etc.

In this context, a major PCN design issue is that of selecting between dynamic TDMA and CDMA based MAC approaches. There are a number of factors to consider, since the choice has a strong impact on system capacity, performance and remote terminal cost/complexity. In general, a properly designed spread spectrum CDMA system can provide system capacity that is ~3-5 times more than that of an equivalent TDMA system. This capacity increase may be attributed to the powerful spatial re-use effect in CDMA, which more than offsets the initial bandwidth penalty due to spreading. CDMA also has the advantage of potentially supporting overlay operation when new frequencies are not available. However, these advantages are offset by several disadvantages: specifically, CDMA requires accurate power control to avoid the "near-far" problem, adding to remote cost/complexity. In addition CDMA does not support high transmission bit-rates, making it less suitable for multimedia applications requiring relatively high transfer rates. On the other hand, TDMA techniques support high burst transmission rates, and can actually be configured to operate with adaptively selected higher efficiency modulation modes to support certain special classes of applications. Also, since the propagation loss constant for PCN with small cells is expected to be ~2 (rather than ~3-4 for cellular mobile), the spatial reuse advantage will not be as large, thus resulting in a CDMA capacity advantage lower than that reported for the Qualcomm digital mobile system proposal, e.g., [GI91]. Our own preliminary results for dynamic TDMA vs. packet CDMA in a mixed voice/data environment show that the CDMA capacity advantage may vary from as little as 25% to as much as 200%, depending on the applicable propagation constant and other factors [WI92]. It is remarked here that there is considerable controversy among experts in the field regarding the relative merits of CDMA and TDMA for PCN; see for example [CO90] and [VI91], and that further work will be required to settle the issue definitively.

3. EXAMPLE MULTIMEDIA PCN ARCHITECTURE: Based on the above discussions, an illustrative example of a possible prototype architecture for PCN is given in Fig. 5. The proposed PCN architecture is based on the ATM-compatible packet switching approach mentioned earlier. The basic unit of data transmitted in the PCN system is a an broadband ISDN/ATM sized cell (packet)

296

containing 48 byte data payload along with appropriate internal network header and trailer. Transmission speeds are estimated to be of the order of 0.25-1.0 Mbps inbound and 1-5 Mbps outbound, with appropriate asymmetrical modulation and coding techniques that minimize remote terminal complexity and power. Medium access control within each microcell is based on a generalized dynamic TDMA approach capable of providing both constant bit-rate (CBR) and variable bit-rate (VBR) transport at various quality-of-service (QOS) levels. Multimedia traffic can be adequately supported within such a framework by employing appropriate dynamic resource allocation techniques for TDMA slot management. In particular, time-of-expiry (TOE) based deadline scheduling features, e.g. [PA88], can be added to the MAC protocol layer to further improve capabilities for handling real-time multimedia traffic [e.g., low bit-rate compressed video]. Data received at microcell base stations is converted to the standard ATM format by removing PCN network headers and replacing these with (5 byte) ATM headers. These ATM data units (cells) are routed among microcells using a hierarchy of SMDS metropolitan area network (MAN) and ATM switching technology (as shown in Fig. 5), providing reasonably good "scalability" as traffic volume and/or the number of microcells grows. The proposed architecture should provide for reasonably seamless multimedia application support over wireline and wireless networks (subject, of course, to peak bit-rate limitations of the PCN).

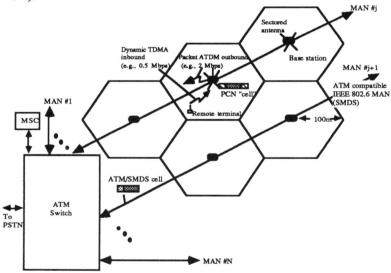

Fig. 5: Example architecture of Multimedia PCN system

In order to obtain additional insight into the proposed multimedia PCN architecture, we focus on the performance of the dynamic TDMA media access control protocol in an integrated voice/data traffic environment. Specifically, a detailed software simulation model has been constructed for a generalized dynamic TDMA media access protocol. The software model itself will not be described here, and the reader is referred to [JO88] for general methodology, and to [WI92] for further specific details about the model.

D-TDMA protocol: The MAC protocol considered for the PCN system architecture outlined above is dynamic packet TDMA, similar to access demand assigned protocols previously proposed for satellite and ground radio applications, e.g. [RA81b]. As discussed earlier, co-existence with future broadband networks is facilitated by the adoption of an ATM-like 48 byte data payload format as the basis for all TDMA transmissions. Fig. 6 illustrates the general frame structure of the dynamic TDMA protocol under consideration.

Frames are sub-divided into request slots and message slots; each frame contains N_r request slots followed by N_t message slots as shown in the figure. Each message slot (after accounting for an appropriate guard time) provides for transmission of a packet or ATM-like "cell" with data payload of 48 bytes and 3 bytes of PCN network header. An additional 2 bytes are allocated in each slot for additional transmission overheads required for modem acquisition, equalizer training, etc. Request slots are comparatively short, and are assumed to contain 6 bytes of control information, in addition to the 2 byte transmission preamble. Of the N_t message slots, a maximum of $N_v < N_t$ slots in each frame can be assigned for connection-oriented CBR voice traffic. Datagram type messages are dynamically assigned one or more 48 byte slots in the TDMA interval following the last allocated voice slot in a frame. Long data messages which cannot be accommodated in a single frame may be segmented for transmission in multiple frames.

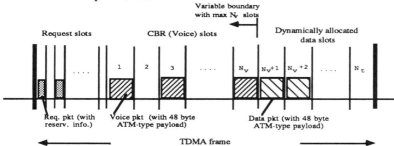

Fig. 6. Frame structure of dynamic TDMA media access control protocol

The basic channel access scheme follows a combination of "circuit mode" reservation of slots over multiple TDMA frames for voice calls, along with dynamic first-come-first-served (FCFS) assignment of remaining capacity for packet data traffic. A request for channel assignment is made in random access (slotted ALOHA) mode using a randomly selected request slot. A terminal which generates either a new voice call or a data packet in the current TDMA frame, randomly selects one of the N_r request slots in the next TDMA frame and transmits an appropriate request packet. Requests which encounter a collision do not receive an acknowledgement and "time-out" before the end of the current TDMA frame. Voice requests are retransmitted in the following frame in a randomly selected request slot, whereas data requests are rescheduled with random delay. For voice calls, requests are repeated until a maximum waiting time for access is reached. At that point, the call is considered "blocked", and a busy tone is returned to the user. Data users do not encounter blocking, and will continue retransmission of requests until an acknowledgement is received from the base station controller. It is noted here that this queued centralized resource allocation framework can readily accommodate service policies other than FCFS. For example, multimedia traffic containing some fraction of time-critical information may benefit from deadline scheduling strategies designed to minimize packet expiry.

Performance evaluation: A direct-event simulation model has been developed for performance evaluation of the D-TDMA access protocol under consideration. This model can be exercised to obtain relevant system performance measures such as channel utilization, voice access delay, voice blocking probability, data delay, etc. The parameters used to obtain the numerical results reported here are summarized in Table 2. The simulation model was exercised over a wide range of loading (0.1 to 0.8 erlangs) on the D-TDMA channel by increasing the number of voice and data users. Also, the mix of voice and data was changed so that voice traffic represents 50%, 63%, 75% or 92% of the total offered traffic.

Table 2: Summary of Performance Evaluation Parameters

TDMA channel speed (R)	240 ; 480 Kbps
TDMA frame length	48 ms
Number of request slots (N_r)	20 (for R=240); 40 (for R=480)
Message slot payload	48 bytes (=ATM)
Request slot payload	6 bytes
Overhead per slot	2 bytes sync. etc. + 6 ms guard time
Number of voice users (@ 8 Kbps each)	U_v
Voice call arrival rate per user	$.0005 \text{ sec}^{-1}$
Average voice call duration	3 min
Max voice call set-up time	5 s
Average request retransmission delay	.04 s
Number of data users	U_d
Average data message length (L)	0.64 ; 5.12 Kb
Data message arrival rate per user	0.1 (for L=0.64); 0.0125 (for L=5.12) sec^{-1}

For a channel speed, R of 240 Kbps and for an average data message length, L of 0.64 Kb performance curves such as channel utilization, voice channel access delay, voice blocking probability and data delay are shown in Figs. 7(a) to (d). Similar plots at channel speeds of 240 Kbps and 480 Kbps and longer data messages with average length of 5.12 Kb, are shown in Figs. 8(a) to (d) and 9(a) to (d) respectively. A key issue is the selection of a value of N_v (voice slot limit) which provides a reasonable balance between voice and data performance. Here, N_v was chosen so that the frame time is divided roughly in proportion to the ratio between offered voice and data traffic. Note that the maximum achievable channel throughput is less than 88% in all cases, since 12% of the TDMA frame is devoted to request slots in this example.

From the channel utilization vs. offered traffic curves in Fig. 7(a), it is observed that with R=240 Kbps and short data messages (640 bits), the channel is uncongested up to about 60% throughput, after which a departure from linearity is observed. From Fig. 7(b), it can be seen that the corresponding voice access time plots begin to curve upwards in the 60-70% throughput region; typically 100 ms access delay can be achieved at the 60% throughput level. As expected, voice access delay depends upon the mix of voice and data traffic, and will tend to increase as the proportion of data traffic rises [note that since slotted ALOHA request slots were engineering to be lightly loaded in all cases, access delay is generally not a significant performance factor].

Fig.7a : TDMA channel utilization versus offered traf

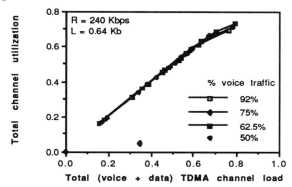

Fig.7b : Voice call access delay versus offered load

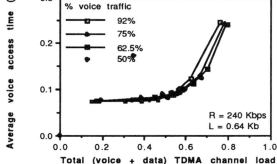

Fig. 7(c) shows the voice blocking probability for the R=240 Kbps and L=640 bit example; in this case, it is observed that voice blocking begins to increase at about 50% channel load, rising quite rapidly as offered load exceeds 55%. As expected, voice blocking increases as the proportion of data traffic is raised. Finally, Fig. 7(d) shows corresponding data delay curves at the four voice/data ratios. The data delay is a crucial performance measure for many multimedia applications, and relatively low values (e.g., less than a few 100's of ms) are desired in many scenarios. It is observed that the data delay starts at a moderately low base value (corresponding to frame latency and transmission time) and then begins to increase as channel load exceeds about 55-60%. Highest data delay is experienced when the proportion of voice traffic is high, as might be expected with most of the TDMA frame occupied by CBR calls.

300

Fig. 7c: Voice call blocking versus offered load

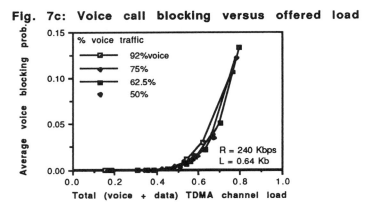

Fig. 7d: Data transmission delay versus offered load

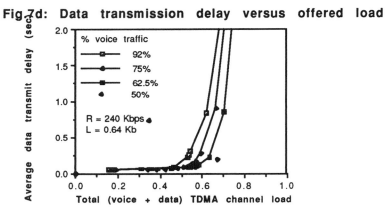

Overall, for this scenario with short data messages, it appears that the system performance (and hence capacity) is generally dominated by voice blocking rather than data delay. Depending on data and voice performance goals, capacities in the region of 50-60% may be achieved for this scenario.

Consider next the results of Fig. 8 for R=240 kbps with "long" 5.12 Kb data messages (as might be the case with file transfer and/or certain image related applications). In this case, it is observed that voice access time curves turn upwards at about 55% load, while voice blocking begins to rise sharply between 45 and 50% load. Data delays in this case also begin to rise at lower loads (e.g., 50%), increasing sharply after about 60% offered traffic. In this case, data delays are higher than those for short messages, due mainly to the longer service time and the associated segmentation for transmission over multiple frames (this is particularly true for the case with the highest proportion of voice traffic). This implies that with long data messages, system performance may be dominated by data delay rather than voice blocking (particularly when strict delay goals are applied).

Fig.8a : TDMA channel utilization versus offered traff

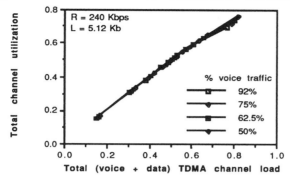

Fig. 8b : Voice access delay versus offered load

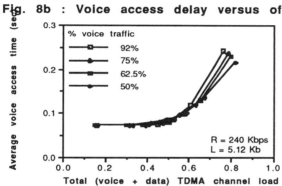

Fig.8c: Voice call blocking versus offered load

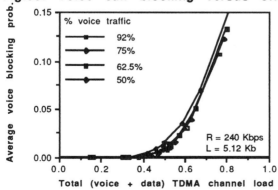

302

Fig. 8d: Data transmission delay versus offered load

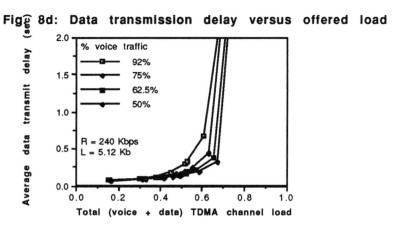

Corresponding results are also obtained for long 5.12 Kb data messages, but with higher channel speed (R=480 Kbps). As expected, performance is generally improved by the higher channel speed. At a given channel utilization level, the improvement in data delay can be quite significant (Fig. 9), especially when stringent delay criteria are imposed.

Fig. 9d : Data transmission delay versus offered load

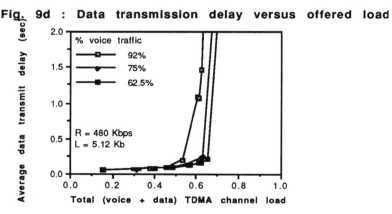

Typical D-TDMA operating points based on a joint performance goal of voice blocking probability (VBP) < 1% and average data delay < 250 ms are summarized in Table 3. In all cases shown, the operating point is determined by the voice blocking requirement (note that this would no longer be the case if delay goals much less than 250 ms are used). For each such 1% VBP operating point, channel utilization, average data delay and "peak" (90th percentile) data delay are given. These peak data delay values (not shown in Figs. 7-9 discussed earlier), are particularly useful for understanding the network's ability to support time-critical traffic that may be generated by multimedia applications.

Table 3: D-TDMA Performance at Operating Points (VBP<1%)

Case	Channel Utilization	Avg. Data Delay (ms)	Peak (90%) Data Delay (ms)
240 Kbps; .64 Kb data			
voice % = 75	0.553	138	257
voice % = 50	0.530	88	166
240 Kbps; 5.12 Kb			
voice % = 75	0.512	190	388
voice % = 50	0.460	122	232
480 Kbps; 5.12 Kb			
voice % = 75	0.621	239	533
voice % = 50	0.596	119	225

The above table's entries for average and peak data delay show that these performance measures are quite sensitive to the ratio of voice:data traffic, and is poorest when voice traffic predominates. This suggests the application of additional measures to maintain data performance. For example, the channel operating point may be intentionally backed off from the above "capacity" levels to improve average and peak delay [this is a particularly viable option for the R=480 Kbps case, since nominal capacities obtained are higher]. Alternatively, one could limit the value of N_V (i.e., the maximum number of slots accessible to voice users) to reserve a somewhat larger fraction of the channel to data. Higher channel speed may also be used as a mechanism for general improvement in the data performance. In view of the importance of peak data delay and/or delay variance ('jitter') in many multimedia applications, it is instructive to consider the distribution of delay in addition to the average.

Based on the peak percentile delay results given in the table above, it can be concluded that the variance of delay could be a significant performance issue in such dynamic TDMA channels. This difficulty (which is fundamental to most PCN channel access procedures) will have to be addressed in order to support future multimedia applications effectively. It is likely that some combination of increased modem speed (e.g., >1 Mbps) and improved protocol operation will be required. Time-of-expiry ("deadline scheduling") techniques are a promising mechanism for providing improved MAC support of time-critical traffic, and will be reported in our future work.

4. CONCLUDING REMARKS: This paper has presented a discussion of system design issues related to multimedia capable PCN systems. We believe that such integrated PCN systems will play an important role in broadband communications networks of the future, and that it is timely to start considering technology selection and system design issues. An example dynamic TDMA based integrated PCN system architecture intended to co-exist with ATM/B-ISDN has been discussed, and some related media access control performance issues have been investigated. Based on our results to date, we believe that acceptable multimedia

304

service can be provided with moderate-to-high speed dynamic TDMA access provided that the system is augmented with additional features for effective support of time critical traffic. Of course, a great deal of further design and prototyping work remains before the technical viability of multimedia PCN systems discussed in this paper can be conclusively established.

5. REFERENCES:

[CO87]   D. C. Cox, "Universal Portable Radio Communications", *Proc. IEEE*, April 1987, pp.436-477.

[CO90]   D. C. Cox, "Personal Communications - A Viewpoint", *IEEE Commun. Mag.*, Nov 1990, pp. 8-20.

[GI91]   K.S. Gilhousen, "On the Capacity of a Cellular CDMA System", *IEEE Trans. on Veh. Tech.*, May 1991.

[GO90]   D. J. Goodman, "Cellular Packet Communications", *IEEE Trans. on Commun.*, Aug. 1990, pp.1272-1280.

[GO91]   D.J. Goodman "Trends in Cellular and Cordless Communications", *IEEE Communications Magazine*, June 1991, pp. 31-40.

[JO88]   K. Joseph and D. Raychaudhuri, "Simulation Models for Performance Evaluation of Satellite Multiple Access Protocols", *IEEE Journal on Selected Areas in Communications*, Jan. 1988, pp. 210-222.

[PA88]   S. Panwar, D. Towsley and J. Wolf, "Optimal Scheduling Policies for a Class of Queues with Customer Deadlines to the Beginning of Service", *J. of the ACM*, Oct. 1988, pp. 832-844.

[RA81a]   D. Raychaudhuri, "Performance Analysis of Random-Access Packet-Switched Code Division Multiple Access Systems", *IEEE Trans. on Communications*, June 1981, pp. 895 - 901.

[RA81b]   S.S. Rappaport and S. Bose, "Demand Assigned Multiple Access Systems Using Collision Type Request Channels: Stability and Delay Considerations", *Proc. IEE*, Vol. 128, No.1, Jan. 1981.

[RA86]   K. Joseph and D. Raychaudhuri, "Stability Analysis of Asynchronous Random Access CDMA Systems", *Proc. IEEE Global Communications Conf.*, Dec 1986, pp. 48.1.1-7.

[SC90]   D.L.Schilling, R.L.Pickholtz and L.B.Milstein, " Spread spectrum goes commercial ", *IEEE Spectrum*, August 1990, pp 40-45.

[ST89]   R. Steele, "The cellular environment of lightweight hand-held portables," *IEEE Communications Magazine*, July 1989, pp. 20-29.

[ST90]   R. Steele, "Deploying Personal Communication Networks", *IEEE Commun. Mag.*, Sept. 1990, pp.12-15.

[VI91]   A.J. Viterbi, "Wireless digital communications: a view based on three lessons learned", *IEEE Commun. Mag.*, Sept. 1991, pp. 33-36

[WI92]   N. Wilson, R. Ganesh, K. Joseph and D. Raychaudhuri, "CDMA vs. Dynamic TDMA for Access Control in an Integrated Voice/Data PCN", to be presented at *1st Intl. Conf. on Universal Personal Communications*, Sept. 1992.

MULTIPLE ACCESS OPTIONS FOR MULTI-MEDIA WIRELESS SYSTEMS

Richard Wyrwas, Weimin Zhang,
M.J. Miller and Rashmin Anjaria
Mobile Communications Research Centre
Signal Processing Research Institute
University of South Australia

Abstract

Alternative access strategies for multi-media wireless systems are discussed. FDMA, TDMA, CDMA and random packet access schemes are considered. A number of possible strategies are presented and the major performance issues are discussed.

1 Introduction

A number of proposals are currently under development for future generation mobile and personal communications systems. These systems will bring with them the potential to provide a wide range of services operating in a wide rang of wireless environments. The term "multi-media" is used to indicate the fact that the information we wish to transmit can come from a variety of sources (voice, facsimile, low and high rate data) with different source characteristics (data rates, activity ratio, burstiness) and different quality (bit error rate) requirements. We also take the term to reflect the fact that we may wish these transmissions to occur in several different channel types (indoor, outdoor, cellular, microcellular) with different transmission characteristics.

It is widely acknowledged that different radio interfaces are required for the different transmission environments that are expected to be encountered. Indeed the pan-European program on Research and development in Advanced Communications in Europe (RACE) suggests four different radio interfaces for the portable terminal for the home, office, vehicular and public environments respectively [1]. However, within each of these radio interfaces will be the need to accommodate sources with different data rates, burstiness and quality requirements. These issues are addressed in this paper.

In this paper we discuss the wireless multiple access options and system performance measures for multi-media wireless systems. We will consider the four main multiple access techniques, namely FDMA, TDMA, CDMA and random packet access.

2 Variable Data Rate Multiple Radio Access

We consider the four main multiple access techniques, namely
- FDMA
- TDMA
- CDMA
- Random Packet Access

For each case, we consider the feasibility of providing for the different message rates that might be required for different services.

2.1 FDMA

In conventional FDMA systems, the total available bandwidth is divided into a number of equi-spaced frequency channels. For reasons of spectral efficiency the transmission rate on a single FDMA channel is usually arranged to be close to the maximum rate. The maximum symbol rate itself is limited by dispersion in the channel. As a result, variable data message bit rates for multi-media applications can only be accommodated by usage of transmission rates less than the maximum rate. This translates directly into loss of spectrum efficiency. A reduced data rate on co-channel links will result in reduced co-channel interference (if the power level is adjusted to keep E_b constant). This will contribute to improved performance in interference limited cellular systems but it cannot be utilised to increase system capacity because the channel allocation and reuse configuration are necessarily fixed (unless sophisticated dynamic channel allocation is employed).

Increased data rates in FDMA systems with identical channels can only be achieved by concatenating a number of channels and this requires either duplicated RF front ends or some form of multi-tone decoding facility. For this reason we will give no further consideration to such systems.

An alternative FDMA system could be conceived with unequal channelisation. A range of channels could be offered depending on the expected source characteristics. For example, a selection of 200kHz, 100kHz, 50kHz, 10kHz, 5kHz and 1kHz channels could be made available for services ranging from video conferencing to message transmission. The system could assign channels of the required data rate as and when required. Such a scheme could achieve stepped variable data rates without recourse to multiple channels. There would be, of course, a number of practical problems with such systems, not least the need to implement mobile terminals with the capacity of different receive and transmit filter bandwidths. However, this problem should be solvable with programmable DSP filter implementations.

Clearly, trunking efficiency may suffer in such a system, and the blocking probability/throughput/delay trade-offs pose an interesting teletraffic problem which are not dealt with here.

2.2 TDMA

TDMA has the advantage over FDMA that the data rate can be adjusted simply by assigning more or less slots (on the same carrier) to a particular user. In mobile radio, systems often referred to as TDMA are, in fact, usually a combination of TDMA and FDMA. For example, the GSM system will typically operate with 3 carriers per cell, each carrier providing 8 TDMA time-slots. Many of the above features applying to variable rate FDMA also apply to TDMA. For example, there is no reason (from a pure resource allocation point of view) why the TDMA carrier bandwidths need to be identical. There are, thus, even more interesting teletraffic problems for the TDMA case.

An "extended TDMA" scheme has been proposed [8]. This is in fact a packet reservation random access type of scheme and will be discussed further in section 2.4.

The conclusion can be drawn that variable bit rates are achievable with TDMA but flexibility is poor.

2.3 CDMA

One claimed advantage for the CDMA scheme is its increased capacity [5] in terms of users/Hz/unit area, assuming proper power control is employed. Capacity measured in this way can be further increased if the user data rate is reduced. In this case the power of each user is assigned according to the required E_b/N_0 for a given bit error level. In this way, a reduced data rate will translate into reduced power being allocated and hence less interference to others. The total capacity measured in data bits/second/Hz/unit area remains the same but the number of users/Hz/unit area is increased.

Both frequency hopping (FH) and direct sequence (DS) CDMA can be employed. For convenience, only DS-CDMA systems are considered here. Forward error control (FEC) schemes are considered as part of the system.

Consider DS-CDMA system with FEC. Let r_b, r_f, r_c represent the user data rate, the FEC coded data rate, and the chip rate respectively and R the FEC coding rate. The information rate r_b is

$$r_b = Rr_f. \tag{1}$$

In a binary CDMA system the processing gain A_p is defined as the spreading ratio

$$A_p = \frac{r_c}{r_f}. \tag{2}$$

The total spreading factor, ψ, is

$$\psi = \frac{r_c}{r_b} = \frac{A_p}{R}. \tag{3}$$

The rates r_b, r_f and r_c are depicted in Figure 1.

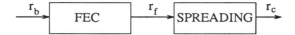

Figure 1: Rate before and after FEC and spreading

Assuming the modulation type is not varied, we will consider 3 obvious combinations of chip rate, processing gain and FEC code rate, which facilitate variable user data rates.

These are the following:

- Fixed processing gain, variable chip rate.
- Fixed chip rate, variable processing gain.
- Fixed chip rate, variable FEC code rate.

Each of these methods has practical implications on the system planning and receiver operations (a Rake type of receiver is assumed).

2.3.1 CDMA with fixed processing gain, variable chip rate

In this scheme, the variable chip rate transmissions occupy variable bandwidths. To maximise capacity it is necessary to ensure that lower data rate users operate with a range of centre frequencies so that not all transmissions are concentrated in the same bandwidth. Thus, for optimum performance, some frequency planning is required.

For different data rate users, the cross-correlation is no longer defined simply by periodical shifts among the codes as in the traditional sense. Even if the pseudo-random codes are taken from a code set with well known cross-correlation properties, it cannot be guaranteed that sequences with different data rates will have low cross-correlation values. Instead, the cross correlation needs to be re-calculated for each code pair with different chip rates. This can make the exhaustive calculation very complex.

This scheme does not appear to be very attractive because the associated unequal bandwidth spreading which complicates frequency planning and RF circuitry.

2.3.2 CDMA with fixed chip rate, variable processing gain

Fixing the chip rate simplifies the receiver RF front end and code synchronisation requirements. However, in this case, variable bit rates translate into variable processing gain. As the bit rate increases the processing gain decreases.

Power control is required in CDMA systems to compensate for the near-far effect. System capacity can fall significantly if accurate control is not provided. Over and above this, power control can be used to allow higher data rates by assigning more power to high data rate users, thus keeping E_b constant. The received power for the kth user will be proportional to its information rate $r_b^{(k)}$ providing equal reliability is required for all users.

More interference will be generated by high data rate users. For given interference conditions the total number of users that can be accommodated at higher data rates will be less than the number that can be accommodated at lower data rates, but the total traffic capacity (in bits per second per Hz per unit area) of the system will remain almost unchanged. In other words, the total system capacity (in bits per second per Hz per unit area) can be arbitrarily shared between any number of users whilst maintaining some given error rate performance objective.

The difference between this system and a system with equal data rate is that a high data rate user will "concentrate" the system interference at a particular location and this may have some repercussions on system capacity.

Good pseudo-random codes with different lengths need to be explored for this application.

2.3.3 CDMA with fixed chip rate and variable FEC code rate

FEC codes with variable coding rate R can be used as a means in spreading the spectrum, alone or along with pseudo-random codes.

It has been suggested by Viterbi [3] that more efficient CDMA can be achieved by using very low rate and powerful convolutional codes in conjunction with pseudo-random codes for scrambling rather than relying on pseudo-random codes to achieve most or all of the spreading. A common cancellation receiver was suggested which involves excessive complexity.

Without a cancellation receiver we can show that there is still a benefit because of the higher coding gain achieved for the convolutional codes with a soft decision maximum likelihood decoder than the processing gain provided by pseudo-random codes. The condition is that a certain signal to noise ratio is required at the FEC decoder which is higher than that required by the despreading process if a pseudo-random code of the equivalent spreading rate is employed. The benefit is achieved by increasing the complexity of decoding since the decoder operates at the chip rate rather than at the bit rate.

With this in mind, it is possible to vary the bit rate for multimedia applications by varying the FEC code rate. This could be achieved by selecting different codes from the same family, or, in the case of convolutional codes, using puncturing. Of course, the different code rates will provide unequal levels of error protection. This may be offset by corresponding power control assignment.

Clearly, there will be complex trade-offs in terms of coding gain (and unequal error protection), processing gain, complexity and power control. The existence of suitable FEC codes is also still an open issue.

A combination of variable rate FEC codes and variable spreading ratios is an alternative solution currently under investigation. Combinations of spread spectrum and TDMA access schemes have also been proposed by some authors [10]. Block FEC codes could also be used for the same purpose, if suitable codes can be found.

Of these three methods for variable data rates, the fixed chip rate schemes are most suitable because they use fixed bandwidth.

2.3.4 Multi-code and multi-level CDMA

In a multi-code CDMA (MC-CDMA) system, a basic data rate r_b is specified according to the user requirement and the channel coherence bandwidth. Higher data rates are achieved by allocating more than one code (channel) to a single user. That is, a message of rate $N_c r_b$ is disassembled (de-multiplexed) into, N_c subsequences each of rate r_b. Then each subsequence is transmitted using one of N_c spreading codes. At the receiver, the N_c received sequences are reassembled (multiplexed) into the required composite message sequence.

This scheme is similar to using multilevel, rather than binary, spreading codes. Under these circumstances it may not be necessary to use multiple, independent receivers. Simplified receiver structures may be possible, for example, making use of fast transforms.

Multi-level CDMA is another topic under investigation. Better pseudo-random codes exist in higher order fields, say $GF(4)$, than in $GF(2)$. For a given number of memory elements there are more codes in higher order fields than in $GF(2)$. This is an ideal property for multi-media applications.

2.4   Random Packet Access

TDMA and Slotted Aloha can be considered as extreme cases of packet access protocols. For TDMA, slots are allocated on a fully reserved basis, and are only available to one particular user for the duration of the transmission. Access overheads are low and slot collisions are completely avoided. However, TDMA is inefficient for bursty traffic because unused slots cannot be reallocated.

On the other hand, Slotted Aloha is a purely random access and any user can contend for access at any time. However, collisions can occur and throughput is poor (limited to only 0.36) under fully loaded conditions. However, in practical systems higher capacity can be achieved due to the capture effect.

A number of compromise solutions have been proposed in recent years. Two of these are Packet Reservation Multiple Access (PRMA) [6, 7] and Extended TDMA [8]. The basic principle behind these schemes is that slots are reserved only when they are needed and are immediately freed-up when the data source stops transmitting (during pauses in speech, for example). This is achieved by marking each slot as either *available* or *reserved*. The system automatically changes the marking from reserved to available when a data source stops transmitting.

The throughput of such packet reservation schemes can be very high (over 0.9) whilst at the same time taking full advantage of the bursty nature of most practical data sources. Performance analysis for such schemes has been presented elsewhere [6, 9, 11].

2.4.1 Variable data rates in random packet access

As with FDMA, the symbol transmission rate associated with each packet will usually be arranged to be close to the maximum rate, which is limited by channel dispersion. There is little to be gained by reducing the transmission rate.

However, variable message rates can be very easily achieved in packet access schemes, simply by using variable numbers of time slots. In fact, such schemes could work particularly well because the system can retain control of packet prioritisation to ensure that resources are allocated on a fair (or at least efficient) basis.

Combined random access/FDMA schemes can be envisaged where data rates significantly higher than the channel coherence bandwidth are required. Alternatively a combination of random access/CDMA could be used.

3 Maximum Date Rates and Data Rate Ranges

For multi-media applications it is important to know the maximum bit rate achievable for the above four access techniques in various channel types.

One common feature for various mobile channels is the multipath effect. It can cause severe inter symbol interference (ISI) and hence restricts the maximum data transmission rate. For the basic receivers, ISI can be almost avoided by ensuring that the symbol rate is less than the coherence bandwidth of the channel. This is useful only for low rate FDMA schemes.

With the use of equalizers, symbol rates of several times the coherence bandwidth are possible for FDMA, TDMA or Random Packet Access systems. A typical outdoor coherence bandwidth can be as low as 50 kHz and in the indoor case it is typically 2

MHz. In the GSM system, for example, the equalizer should be capable of operating with a delay spread of 4 symbol periods.

For CDMA we assume Rake receivers will be used which combines signals from several paths. The time-varying property of the multipath channel can be estimated in a better way in a CDMA receiver than in FDMA and TDMA due to the wider bandwidth (more information) used. For the simplest Rake receivers, the maximum bit rate is still restricted by delay spread. However, for slightly more sophisticated receivers bit rates much higher than the channel coherence bandwidth may be achieved [2].

In future micro cellular situations there is likely to be a strong line of sight signal component and consequently maximum data rates in the order of Mbps should be possible.

There have been many proposals recently for the application of more spectrally efficient modulation schemes, such as 16-QAM, for the mobile radio channel [13]. There is some debate as to whether this provides any real gain in frequency reuse situations, because the increase in required carrier to interference ratio results in an increase in frequency reuse distance which more than offsets the modulation improvement [12].

For FDMA, TDMA and packet access systems, DSP and programmable filter technology has made possible modems capable of handling a wide range of data rates, with the limitation being the speed and complexity of the programmable devices. However, the coherence bandwidth of the communications channel provides a more fundamental limitation to the maximum data rate in a single link. With this restriction, different data rates can be achieved by concatenation of channels and/or time-slots. This implies some additional hardware complexity, especially in the case of FDMA.

4   Source Burstiness

Where the source is bursty it is desirable to make the channel resource available to other users during silence periods, or at the very least to ensure that interference generated by the silence is minimised.

With FDMA and TDMA systems it is easy to detect silence periods and effectively switch off the transmitter at these times. However, in general it is not possible to reallocate the channel to another user until the transmission is completely terminated.

One significant advantage of both CDMA and packet access schemes is that they allow the channel resource to be reallocated to other users. In the case of CDMA this is achieved indirectly by the fact that the reduced interference allows other users (with different codes) to access the system and still maintain an acceptable interference level (this is sometimes referred to as statistical multiplexing). For packet access techniques the channel is directly made available to other users while a particular user is silent.

In both cases there are overheads, however. In the case of CDMA a minimum transmission bit rate greater than zero is often required to maintain synchronisation during silence periods. In the case of packet access the silent user has to re-contend for the channel when the silence period ends.

5 Transmission Quality Requirements

In all multiple access schemes transmission quality in any given channel and transmitter quality can be traded (until the error floor is reached). Similarly, channel coding and, in the case of CDMA, processing gain, can be used to adjust bit error rate performance. In the latter two cases some bandwidth adjustment will be required.

Adjustment of transmitter power to achieve the desired transmission quality has serious implications for frequency reuse (cellular) schemes with a pre-assigned frequency plan (most FDMA, TDMA cellular systems) because interference levels are changed and the given reuse distances may no longer be sufficient. In the case of CDMA, adjustment of transmitter power for a given user does impact on overall system capacity, but since this capacity has a soft limit, significant disruption is less likely.

6 Summary

Four alternative multiple access strategies for multi-media wireless access have been discussed. In the case of CDMA, three possible combinations of processing gain, chip rate and FEC code rate have been considered. Comparisons of respective performance levels in terms of amenability to variable bit rates, variable source characteristics and variable quality requirements have been made.

7 Acknowledgements

The Mobile Communications Research Centre is funded by Australian Overseas Telecommunications Corporation (AOTC). Helpful comments from AOTC Research Laboratories staff were much appreciated.

References

[1] Evci, C.C., *Pan-European project for third generation wireless communications*, Third Winlab Workshop on Third Generation Wireless Information Networks, p 19-39, April 1992.

[2] Turin, G. L., *Introduction to spread-spectrum antimultipath techniques and their application to urban digital radio*, Proc. IEEE, Vol. 68, No. 3, p 328-353, March 1980.

[3] Viterbi, A. J., *Very low rate convolutional codes for maximum theoretical performance of spread spectrum multiple access channels*, JSAC-8, p641-649, May 1990.

[4] Viterbi, A. J., *Wireless digital communication: A view based on three lessons learned*, IEEE Comms. Mag., p33-36, Sep. 1991.

[5] Gilhousen, K. S. et al., *On the capacity of a cellular CDMA system*, IEEE Trans. Vehicular Tech. Vol. 40, No. 2, May 1991.

[6] Goodman, D. J. et al., *Packet reservation multiple access for local wireless communications*, Proc. 38th IEEE V. Tech. Conf., Philadelphia, p701-706, June 1988.

[7] Goodman, D. J., *Cellular packet communications* IEEE Trans. Comm., Vol. 38, No. 8, p 1272-1280, August 1990

[8] *GM, Hughes enter digital cellular market with new extended-TDMA technology*, Telecommunications Reports, p14, Dec. 17 1990.

[9] Sheikh, U. H. A. et el., *The ALOHA systems in shadowed mobile radio channels with slow or fast fading*, IEEE Trans. V. Tech., Vol.39, No.4, p289-297, Nov. 1990.

[10] Mizuike, T. et al., *Burst scheduling algorithms for SS/TDMA systems*, IEEE Trans. Comm., Vol. 39, No. 4, p533-539, April 1991.

[11] Wang, X. et al., *Performance analysis of a combined random-reservation access scheme*, IEEE Trans. Comm., Vol. 39, No. 4, p478-481, April 1991.

[12] Cox, D. C., *Universal digital portable radio communications*, Proc. IEEE, Vol. 75, No. 4, p436-477, April 1987.

[13] Webb, W. et al., *Does 16-QAM provide an alternative to the half-rate GSM speech modem*, Proc. 41st IEEE V. Tech. Conf., St. Louis, p511-516, May 1991.

Optimal Code Rates for CDMA Packet Communications with Convolutional Coding

Brian D. Woerner[1], Assistant Professor
The Bradley Department of Electrical Engineering
Virginia Polytechnic Institute and State University
Blacksburg, VA 24061-0111

Abstract

We consider a direct-sequence spread-spectrum multiple access communication system with convolutional error correction coding and hard decision decoding. This system operates as a wireless packet communications network. We explore several techniques to calculate the packet error probability and throughput of such a system. These techniques can be clasified as either bounding methods or approximations. We describe a bounding technique which improves on the best previous bounds and compare this bound to a recent approximation. Although the two techniques yield very different values for throughput, both techniques result in similar choices for the optimum rate of an error correction code.

1. Introduction

Commercial applications have begun to blossom for code division multiple access (CDMA) techniques based on direct-sequence spread-spectrum (DS/SS) technology. In recent years, there has been strong interest in using CDMA techniques for packet communications within wireless information networks. Several researchers have proposed CDMA technology for indoor wireless LANs. The performance measures of interest for a CDMA packet radio network are packet error probability and throughput. Throughput can be computed directly from packet error probability. In this paper, we consider a CDMA system with convolutional coding. We examine the problem of selecting the code rate for a convolutional code which will maximize system throughput. We obtain results using both exact bounds and approximate techniques, and then compare these results.

1.1. Previous Work

The problem of computing packet error probability can be broken into two parts. First, we must compute the bit error probability and then use our result for bit error probability to compute packet error probability. The difficulty of this process is that bit error events within a packet are correlated with one another [4]. Two approaches have been followed to circumvent the problem of dependent bit error events within a packet: approximation and bounding techniques.

---

[1] The author is with the Mobile and Portable Radio Group at Virginia Tech

320

One simple approximation is to assume that bit error events occur independently, ignoring the dependence of bit errors within a packet. The authors of [2] make this approximation to evaluate the throughput. They report results for the case of an unslotted packet radio network with block coding.

A second approach is to obtain bounds on packet error probability. In [8], Pursley and Taipale obtain an upper bound on bit error probability by considering the worst case of exactly synchronous multiple access interference. This loose upper bound on bit error probability enables them to neglect the dependence of the bit error events, and obtain an upper bound on packet error probability of a slotted packet radio system with convolutional coding. This is approach also followed [11] to compute bounds on the performance of unslotted systems with convolutional coding. The authors of [10] use this technique to draw comparisons between the performance of systems with block and convolutional error correction codes.

In [4], Morrow and Lehnert introduce a fundamentally new approach the problem of computing packet error probability and throughput. They account for the dependence of bit error events by making use of the theory of moment spaces [1]. As a result, they are able to report accurate approximations for the packet error probability and throughput for the case of slotted systems with block codes. The authors of [9] have demonstrated a technique for simplifying the calculations of [4].

In [14] and [13], we used moment space techniques to compute bounds on the packet packet error probability and throughput of both slotted and unslotted CDMA systems with convolutional coding. We extended the work of [4] in several significant ways: we considered convolutional coding, the results we reported were true bounds rather than approximations, and we considered both the slotted and unslotted cases. Additionally, we demonstrated an improved Union/Chernoff bound on error event probability of a convolutional code which also led to an improved bound bound. The resulting bound on packet error probability improved on the bounds of [8] and [11] by approximately one order of magnitude, for the cases of slotted and unslotted systems respectively.

1.2. Organization of this paper

In this presentation, we compute bounds on the performance of a CDMA packet radio system with convolutional coding. We report results for both packet error probability and throughput. Focusing on the case of slotted systems, we determine the optimal code rates which maximize system throughput. We then repeat the calculations using the approximation techniques developed in [9], again determining optimal code rate. We compare the results from bounding techniques with the results obtained from approximate techniques. We conclude that although the bounding techniques and the approximations result in very different throughputs, the choice of optimal coding rate is relatively independent of the choice of performance analysis technique.

2. System Model

In this section, we describe the model used in this paper for a DS/SS system with multiple-access interference. We also review the techniques available for computing bit error probability, and describe simple traffic models for slotted and unslotted packet radio systems.

2.1. A DS/SS MA System

We consider a DS/SS MA system with binary phase shift keyed (BPSK) signalling, and a correlation receiver. Our model is based on [6]. There are a total of K users transmitting over a common channel. Associated with each user $k \in \{1, \ldots, K\}$ is a data signal $b_k(t)$ and a signature waveform $a_k(t)$ which are functions of time. These are defined by

$$b_k(t) = \sum_{i=-\infty}^{\infty} b_{k,i} \psi_T(t - iT) \tag{1}$$

$$a_k(t) = \sum_{j=-\infty}^{\infty} a_{k,j} \psi_{T_c}(t - jT_c), \tag{2}$$

where $\{b_{k,i} \in \{+1, -1\}\}$ is an infinite sequence of encoded data bits, $\{a_{k,j}\}$ is an infinite random signature sequence with each chip $a_{k,j}$ independent and equiprobably distributed on $\{+1, -1\}$, and $\psi_T(\cdot)$ is the unit pulse function of duration T, defined by

$$\psi_T(t) = \begin{cases} 1, & t \in [0, T) \\ 0, & \text{else.} \end{cases} \tag{3}$$

The duration of each encoded data bit is T, while the duration of each chip in the signature signal is T_c. As a result the number of chips per bit is $N = T/T_c$, where N is a integer.

Each user generates a signal $s_k(t)$ by modulating the data signal by its signature signal and a carrier waveform, with the result

$$s_k(t) = \sqrt{2P} \cos(\omega_c t + \phi_k) a_k(t) b_k(t), \tag{4}$$

where P is the signal power, ω_c is the carrier frequency, and ϕ_k is a random phase, uniformly distributed on the interval $[0, 2\pi)$.

A correlation receiver receives the signal r(t) which is the sum of delayed versions of all transmitted signals and thermal noise. The received signal r(t) is

$$r(t) = n(t) + \sum_{k=1}^{K} s_k(t - \tau_k), \tag{5}$$

where $n(t)$ is a white Gaussian process with two-sided power spectral density $N_o/2$, and τ_k is a random delay, uniformly distributed on $[0, T)$. The synchronous correlation receiver recovers the transmitted data bit by correlating $r(t)$ with the transmitter to form a decision statistic $Z_{k,i}$ where

$$Z_{k,i} = \int_{\tau_k}^{T+\tau_k} r(t)\cos[\omega_c(t-\tau_k)+\phi_k]a_k(t-\tau_k)dt. \tag{6}$$

The decision statistic is used to form an estimate $\hat{b}_{k,i}$ of the data bit $b_{k,i}$ based on the rule

$$\hat{b}_{k,i} = \begin{cases} 1, & Z_{k,i} \geq 0 \\ -1, & Z_{k,i} < 0 \end{cases}. \tag{7}$$

A bit error occurs if $\hat{b}_{k,i} \neq b_{k,i}$. We consider a system in which a convolutional encoder with rate r is used in conjunction with hard decision Viterbi decoding. Although it is possible to obtain better performance with soft-decisions, hard-decision decoding is commonly used in wireless communications because it is simpler to implement, particularly with a time varying channel.

2.2. Packet Transmission Systems

In a packet transmission systems, blocks of data are grouped into blocks of consecutive bits called packets. We assume that one packet has a duration of L information bits. A single uncorrected error forces retransmission of the entire packet of data.

We classify a packet transmission system as either *slotted* or *unslotted*, depending on whether all active users transmit packets simultaneously during time slots of fixed duration. In an slotted system, the number of users K is fixed over the duration of the entire packet. The length L of all packets is a constant. Although the users are coarsely synchronized in that they commence and cease transmission at approximately the same time, they are not finely synchronized. The relative phases $\{\theta_k\}$ and delays $\{\tau_k\}$ of the interfering users with respect to the signal of user 1 may be modeled as a collection of independent random variables, unformly distributed on the intervals $[0,\pi)$ and $[0,T)$ respectively. Note that although the phases $\{\theta_k\}$ and delays $\{\tau_k\}$ are initially random variables, they may remain relatively constant over a significant portion of the packet. As a result, the multiple access interference during a bit is significantly correlated with the multiple access interference during the previous bit. For a slotted system, we assume an infinite population of potential users exists, and that the number of users K which transmit during a given slot is a Poisson random variable with probability mass function $p_K(k)$ given by

$$p_K(k) = \frac{G^k \exp(-G)}{k!}, \qquad k = 0, 1, \ldots. \tag{8}$$

The expected number of users G is called the offered traffic.

In an unslotted system, users commence and cease transmission independently, and the number of interferers can vary during the transmission of a packet. The length L of the packet may also vary in an unslotted system. We will focus primarily on slotted systems for ease of explaination. However, the important results can be extended to unslotted systems as well.

2.3. Performance Measures

The most basic performance measure of a communication system is the bit error rate. If K users share the common channel, then the average probability of bit error $\bar{\rho}_K$ is defined by

$$\bar{\rho}_K = \Pr\left[\hat{b}_{k,i} \neq b_{k,i}\right].$$ (9)

It is convenient to express the probability of bit error as a function of the delays $\tau = \{\tau_1, \ldots, \tau_K\}$ and phases $\phi = \{\phi_1, \ldots, \phi_K\}$. We express this conditional probability as $\rho(\tau, \phi) = \Pr[\hat{b}_{k,i} \neq b_{k,i} \mid \tau, \phi]$. Note that the random processes $n(t)$ and $a_K(t)$ are independent from one data interval to the next. Therefore, when conditioned on the delays τ and phases ϕ, all bit error events are nearly independent of one another.[2] It is shown in [8] that when all users have equal signal power the worst bit error rate occurs when all multiple-access interferers are phase and chip synchronous ($\tau = 0$ and $\phi = 0$) with the desired signal. We denote the bound on $\bar{\rho}_K(\tau, \phi)$ obtained from this fact as ρ_K^U and write

$$\rho_K(\tau, \phi) \leq \rho_K^U = \rho_K(0, 0), \qquad \forall \tau, \phi.$$ (10)

If it is phase and chip synchronous, the multiple access inteference has a binomial probability distribution and ρ_K^U can be computed from the formula

$$\rho_K^U = \sum_{j=0}^{(K-1)N} \binom{(K-1)N}{j} 2^{(1-K)N} q(j),$$ (11)

where $q(j)$ is given by

$$q(j) = Q\left(\sqrt{\frac{2rPT}{N_0}}\left[1 + \frac{2j - (K-1)N}{N}\right]\right),$$ (12)

and $Q(\cdot)$ denotes the standard Q-function. The rate of the error correction coder is r. Since the bound ρ_K^U is valid for all τ and ϕ, this bound can be used to avoid the issue of how $\rho_K(\tau, \phi)$ depends on the phases and delays. This approach is followed in [8] and [11]. The true average bit error probability, however, is found by taking the expected value of $\rho_K(\tau, \phi)$ over these phases and delays

$$\bar{\rho}_K = E_{\tau, \phi}\left[\rho_K(\tau, \phi)\right].$$ (13)

---

[2] There will be one chip $a_{k,j}$ from each interferer which overlaps two adjacent bits $b_{1,i}$ and $b_{1,i+1}$. For $K = 2$, the bit error events are still independent, however, for $K > 2$ adjacent bit error events are not strictly independent, although this is a very minor approximation

In [3], a technique is presented for efficiently evaluting the multidimensional integral inplied by (13) to an arbitrary level of accuracy. There may be a large gap between the bit error rates obtained from (11) and (13).

Error Event Probability

Next, we define the probability of failure of the convolutional code. An error event is said to occur if the Viterbi decoder selects a path through the code trellis which diverges from the correct path beginning in the current interval. We denote the probability of an error event in a hard-decision decoded system as $P_\mu(\rho)$ to indicate its dependence on the encoded bit error rate ρ. The error event probability $P_\mu(\rho)$ may be upper bounded by the following expression due to Van de Meerberg [12]. For a memoryless channel with bit error probability ρ,

$$P_\mu(\rho) \;\leq\; \Gamma_{n_o} \left\{ \frac{1}{2}\left[T(D)+T(-D)\right] + \frac{1}{2}D\left[T(D)-T(-D)\right]\right\}_{D=2\sqrt{\rho}} \tag{14}$$

where

$$\Gamma_{n_o} = \left(\begin{array}{c} 2n_o - 1 \\ n_o \end{array} \right) 2^{-2n_o}, \tag{15}$$

$T(D)$ is the transfer function of the convolutional code, and n_o is related to the free distance of the code by the expression $n_o = \lfloor \frac{d_{\text{free}}+1}{2} \rfloor$.

A slightly tighter upper bound may be given by

$$P_\mu(\rho) \leq \sum_{i=d_{\text{free}}}^{M} t_i(P_i - D^i) + T(D) \Bigg|_{D=2\sqrt{\rho(1-\rho)}}, \tag{16}$$

where

$$P_i = \begin{cases} \sum_{j=\frac{i+1}{2}}^{i} \left(\begin{array}{c} i \\ j \end{array} \right) \rho^j (1-\rho)^{i-j}, & n \text{ odd} \\[2mm] \sum_{j=\frac{i}{2}+1}^{i} \left(\begin{array}{c} i \\ j \end{array} \right) \rho^j (1-\rho)^{i-j} + \frac{1}{2} \left(\begin{array}{c} i \\ \frac{i}{2} \end{array} \right) \rho^{\frac{i}{2}}(1-\rho)^{\frac{i}{2}}, & n \text{ even} . \end{cases} \tag{17}$$

t_i is the ith coefficient of the transfer function, and M is a large integer. An algorithm for computing the transfer function in closed form is given in [5]. The proof of this Improved Union-Chernoff Bound is contained in [13].

Packet Error Probability

A single error event causes a packet error to occur. We define the packet error probability as P_E and it can be expressed as

$$P_E = \Pr[1 \text{ or more error events during signaling intervals } 1, \ldots, L]. \tag{18}$$

For a slotted system, we shall find it necessary to condition P_E on the number of users K sharing the system, writing $P_E(K)$ for the packet error probability given that K users transmit during the slot.

Throughput

Throughput is a measure of the total amount of information transmitted by a system which is closely related to packet error probability. A packet is successful if, after error correction by the Viterbi decoder, there are no errors in the packet. Since only successful packets contribute to throughput, we determine throughput by taking the expected number of successful packet transmissions. The normalized throughput S of a slotted system is defined as

$$S = \frac{r}{N} \sum_{k=1}^{\infty} p_K(k)k[1 - P_E(k)].$$ (19)

where S is normalized for the rate r of the convolutional code, and bandwidth expansion N of the DS/SS system. In this treatement, we ignore retransmissions of incorrect packets.

3. Throughput of a Slotted System.

In this section we discuss several techniques for bounding the throughput of a slotted system.

3.1. Upper Bound Based on Synchronous Interference

If there are L bits in a packet, Pursley and Taipale [8] show that the packet error probability $P_E(K)$ for a K user system is upper bounded by

$$P_E(K) \leq 1 - [1 - P_\mu(\rho_K(\tau,\phi))]^L ,$$ (20)

where $P_\mu(\rho)$ is the error event probability for the convolutionally coded system

In [8], the upper bound of (11) is used to evaluate (20), yielding the upper bound

$$P_E(K) \leq P_{E1}(K) = 1 - [1 - P_\mu(\rho_K^U)]^L .$$ (21)

Since phases τ and delays ϕ are correlated from one interval to the next in some unknown manner, the bound of (21) assumes the worst case for τ and ϕ. This assumption is quite pessimistic. We denote the throughput obtained by substituting $P_{E1}(K)$ into (19) as S_1.

3.2. Upper Bound Based on Moment Spaces

There may be a large gap between the bit error rate computed from (11) and $\bar{\rho}_K$ computed from (13). We exploit this gap to improve upon the bound P_{E1}. The principal result follows from the Moment Space Theorem [1]. This theorem states that if two random variables are related by a function, then the expected value of one of those random variables must lie in the intersection between the convex hull of that function and the expected value of the other random variable. Morrow and Lehnert [4] first applied this idea to study of the packet error probability of slotted systems with block coding.

The idea of a convex hull can aid in relating bit error probability to packet error probability. Both $\rho(\tau,\phi)$ and P_E are random variables, and the packet error probability P_E is some function $f(\rho(\tau, \phi))$. The Moment Space Theorem tells us that the expected value of P_E lies within the intersection of the convex hull generated by $f(\rho(\tau, \phi))$ and the expected value of $\rho(\tau, \phi)$. If we use the upper bound ρ_K^U on bit error probability, we get the upper bound discussed in the previous section. However, by examining the intersection of the condition $\rho = \bar{\rho}_K$ with the convex hull, we can arrive at a new upper bound on packet error probability P_{E2}.

The new, tighter upper bound on $P_E(K)$ is given by the following result.

Proposition 1: *Consider a slotted DS/SS MA system with K users and let c be a real number such that*

$$f(\rho) = 1 - [1 - P_\mu(\rho)]^L \qquad (22)$$

is a convex function of ρ on the range $[0, c]$. Then packet error probability may be upper bounded by

$$P_E(K) \le P_{E2}(K) = \begin{cases} \frac{\bar{\rho}_K}{\rho_K} P_{E1}(K), & \rho_K^U \in [0, c) \\ 1, & \rho_K^U \notin [0, c). \end{cases} \qquad (23)$$

The proof of this proposition is contained in [13] and [15].

We call the resulting bound on packet error probability $P_{E2}(K)$. The restriction that $f(\rho)$ be a convex function for $\rho \in [0, c)$ is a mild one. In most cases $f(\rho)$ is convex for all ρ of interest. Normalized throughput for the slotted case is given by (19). Truncating the summation in (19) results in a lower bound on achievable throughput, as does using an upper bound on $P_E(k)$. We call the bounds on throughput obtained by using $P_{E1}(K)$ and $P_{E2}(K)$, S_1 and S_2 respectively. In [14], the bound $P_{E1}(K)$ is compared with the new bound $P_{E2}(K)$ and and S_1 is compared with S_2. It was observed that $P_{E2}(K)$ is an order of magnitude less than the $P_{E1}(K)$ under a variety of channel loading conditions. As a result, the throughput S_2 was significantly greater than S_1, particularly in the heavily loaded case.

3.3. Approximation Techniques

Although the technique described in the previous section yields a true upper bound on packet error probability, and a lower bound on the throughput which can be achieved, the required calculations are quite computationally intensive. As a result, several researchers have investigated the use of simpler approximations. Morrow and Lehnert [4] considered an "improved Gaussian approximation" for the multiple access interference in which the multiple access interference is modeled as a Gaussian random variable conditioned on the variance Ψ of the multiple access interference itself. An even simpler approximation was suggested by Simpson and Holtzman [9]. Applying their result, we find that the packet error probability of a DS/SS system with convolutional coding may be approximated by

$$
P_E(K) \approx P_{E3}(K) \;=\; \frac{2}{3} g \left(Q \left[\frac{N}{\sqrt{\mu}} \right], L \right) + \frac{1}{6} g \left(Q \left[\frac{N}{\sqrt{\mu - \sqrt{3}\sigma}} \right], L \right)
$$
$$
+ \frac{1}{6} g \left(Q \left[\frac{N}{\sqrt{\mu + \sqrt{3}\sigma}} \right], L \right) \tag{24}
$$

where the function $g(\rho, L)$ is given by

$$
g(\rho, L) = 1 - [1 - P_\mu(\rho)]^L , \tag{25}
$$

and where μ and σ^2 are given by

$$
\mu \;=\; \frac{N^2 N_o}{2rPT} + (K - 1)\frac{N}{3} \tag{26}
$$
$$
\sigma^2 \;=\; (K - 1)\left[\frac{23}{360}N^2 + N \left(\frac{1}{20} + \frac{K-2}{36} \right) - \frac{1}{20} - \frac{K-2}{36} \right]. \tag{27}
$$

Note that here μ is taken to be the expected value of the variance of the interference *plus* the channel noise. Using this approximation evaluate the packet error probability, we can compute an approximate throughput S_3.

3.4. Throughput of an Unslotted System.

The analysis of unslotted systems is complicated by the fact that the number of interferers varies with time throughout the duration of a packet. Storey and Tobagi [11] show that the bound of (21) can be extended to the case of unslotted system by use of an auxiliary Markov chain. In this treatment, we focus primarily on slotted systems. However, we have also derived an improved upper bound on packet error probability for the the case of unslotted system. This bound improves the upper bound on packet error probability of [11] by roughly an order of magnitude. A proof of this proposition is contained in [13] and [15].

4. Selection of Optimum Throughput and Code Rates

In order to explore the relationship between the choice of error correcting code and system performance, we used the techniques of the previous section to evaluate the throughputs S_2 and S_3 for a large number of different code rates. For the results reported here, we examined a DS/SS MA system with processing gain $N = 63$ and the number of users ranging from $K = 0$ through $K = 60$. Results are reported here for $E_b/N_o = 8$ dB. We also assume that each packet consists of 1000 information bits to which coding is added.

We examined convolutional codes with rates ranging from $r = 1/8$ to $2/3$. In order to insure that the codes were of comparable complexity, we held the number of states times the number of branches emerging from each state constant for each code. A rate 1/2 constraint length 6 code served as the baseline system.

Figures 1 and 2 plot the throughput S versus offered traffic G for both the lower bound on throughput and the approximation technique, respectively. For each different code, the throughput increases to some maximum level as G increases, and then decays once the error correcting capability of the code is exceeded. The peak throughput occurs at a higher traffic level for the more powerful (lower rate) codes.

With one exception, the lower bound reports smaller values for throughput than the approximation technique. The difference in reported throughput becomes quite large for the condition of high offered load. In practice, this large throughput is obtained at the price of frequent packet errors and retransmissions. Figure 3 plots the throughput maximum throughput $S^*(r)$ of any system with code rate r. Note that both techniques predict that the system should attain its maximum level of throughput when $r = 1/4$. In the case of $r = 2/3$ the approximation is actually less than the lower bound. This is a result of either error in the approximation or a violation of the convexity conditions required by Proposition 1.

Figure 4 plots throughput versus offered traffic, assuming that in each case, the code rate r is selected to optimize system performance for the value of G. For low levels of offered traffic, the lower bound and approximation track one another closely. For large G, the approximation predicts dramatically larger throughput levels. However, the optimum codes rates $r^*(G)$ predicted by the two techniques for a given G remain remarkably close to one another throughout the entire range of G. These optimum code rates are displayed in Table 1. As G increases, the predicted optimum codes rates are reasonably close, even though the predicted throughput is dramatically different. The lower bound method suggests only a slightly more conservative choice for large G. This implies that the system performance is relatively robust with respect to the technique used to select the code rate.

5. Conclusions

In this paper we have investigated several techniques to compute the through-put of a CDMA packet system with convolutional coding. These techniques include both improved lower bounds on throughput and an approximation. The two methods lead to widely divergent values for normalized through-put, particularly under heavy traffic conditions. However, when the code rate r is selected to optimize the normalized throughput, both the lower bound and the approximation technique lead us to choose similar values for r. Under heavy traffic conditions, the lower bounding technique results in a slightly more conservative choice for code rate. We conclude that the simpler approximation technique may be used as to determine code rate in system design. Further investigation is required to compare the accuracy of the two methods in computing values for throughput.

Acknowledgements

The author would like to thank the Unisys Corporation and the Mobile and Portable Research Group's Industrial Affiliates for their support of this work. The author would also like to that Joseph Lichtenstein for his assistance in generating numerical results.

References

[1] M. Dresher, "Moment spaces and inequalities," *Duke Math. Journal*, vol. 20, pp. 261-271, June 1953.

[2] K. Joseph and D. Raychaudhuri, "Performance evaluation of assynchronous random access CDMA with block FEC coding," *Proc. 1989 ICC*, pp. 41.3.1-41.3.5, 1989.

[3] J. S. Lehnert and M. B. Pursley, "Error Probabilities for Binary Direct-Sequence Spread-Spectrum Communication with Random Signature Sequences," *IEEE Trans. Commun.*, vol. COM-35, pp. 87-98, January 1987.

[4] R. K. Morrow and J. S. Lehnert, "Bit-to-bit error dependence in slotted DS/SSMA packet systems with random signature sequences," *IEEE Trans. Commun.*, vol. COM-37, pp. 1052-1061, October 1989.

[5] I. M. Onyszchuk, "On The Performance of Convolutional Codes," *Ph.D. Thesis*, California Institute of Technology, 1990.

[6] M. B. Pursley, "Spread Spectrum Multiple Access Communications," Multi-User Communication Systems, New York: Springer-Verlag, pp. 139-199, 1981.

[7] M. B. Pursley, D. V. Sarwate and W. E. Stark, "Error probability for direct-sequence spread-spectrum multiple access communications - Part I: Upper and lower bounds," *IEEE Trans. Commun.*, vol. COM-30, pp. 975-984, May 1982.

[8] M. B. Pursley and D. J. Taipale, "Error probabilities for spread-spectrum packet radio with convolutional codes and Viterbi decoding," *IEEE Trans. Commun.*, vol. COM-35, pp. 1-12, 1987.

[9] F. Simpson and J. Holtzman, "A DS/CDMA packet error probability approximation," *Proceedings of the 29th Allerton Conference on Communication, Control, and Computing*, Urbana, IL, October 1991, pp. 352-361.

[10] B. Smith, M. Georgiopoulos, M. Belkerdid, "Comparison of BCH and convolutional codes in a direct sequence spread spectrum multiple access packet radio network," *Proceedings 1991 MILCOM*, pp. 43.5.1-43.5.5, Maclean, VA, Nov. 1991.

[11] J. S. Storey and F. A. Tobagi, "Throughput performance of an unslotted direct-sequence SSMA packet radio network," *IEEE Trans. Commun.*, vol. COM-37, pp. 814-823, August 1989.

[12] L. Van de Meerberg, "A tightened upper bound on the error probability on binary convolutional codes with Viterbi decoding," *IEEE Trans. Info. Theory*, vol. IT-20, pp. 389-391, May 1974.

[13] B. D. Woerner, "The Application of Coded Modulation to DS/SS Communications," *Ph.D. Thesis*, University of Michigan, 1991.

[14] B. D. Woerner and W. E. Stark, "Packet error probability of DS/SSMA Communications with convolutional codes," *Proceedings 1991 MILCOM*, pp. 6.1.1-6.1.5, Maclean, VA, Nov. 1991.

[15] B. D. Woerner and W. E. Stark, "Improved bounds on the performance of DS/SS with convlutional codes," in preparation for *IEEE Trans. Commun.*

Table 1: Optimal choice of code rates predicted by the two techniques for selected values of G.

G	$r*(G)$ predicted by bound	$r*(G)$ predicted by approximation
1.0	2/3	1/2
2.0	2/3	1/2
5.0	1/2	1/2
10.0	2/5	2/5
15.0	1/3	1/3
20.0	1/4	1/3
25.0	1/5	1/4
30.0	1/6	1/4
35.0	1/8	1/5
40.0	1/8	1/6

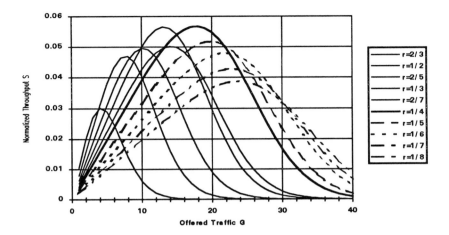

Figure 1: Normalized throughput S_2 versus offered traffic G for several code rates using the lower bounding technique.

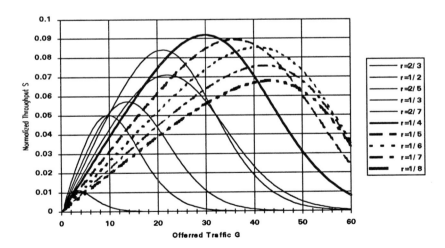

Figure 2: Normalized throughput S_3 versus offered traffic G for several code rates using the approximation technique.

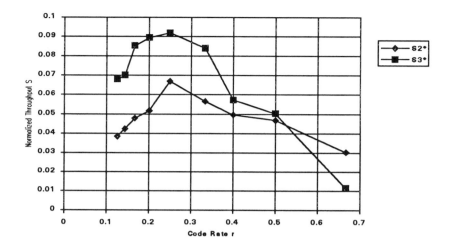

Figure 3: Maximum normalized throughput $S^*(r)$ as a function of rate r for both systems.

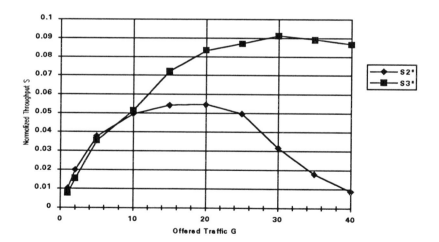

Figure 4: Maximum normalized throughput using rate $r^*(G)$ as a function of offered traffic G for both systems.

INDEX

A